THE HUMAN REALITY

Valentin Matcas, M.Ed.

Copyright © 2026 Valentin Matcas

All rights reserved.

ISBN: 9781973474265

DEDICATION

I dedicate this book to everyone eager to learn and develop continuously throughout life.

CONTENTS

1. The Human Reality, Contorted — 1
2. The Real and the Consensual Human Reality — 7
3. Accurate and Consensual Truth, the Duality of the Human Reality — 41
4. Consensual Theories and the Entire Scientific Consensus — 69
5. Cosmology, Cosmogony, and the Everlasting Brotherhood — 86
6. Old Religions, New Religions, and the Actual Divine — 149
7. The Big Bang and the Entire Consensual Mechanism Behind — 211
8. Accurate Model of the Human Reality — 250

1 THE HUMAN REALITY, CONTORTED

There is a difference between the material objective world and the actual human reality, since the human reality is very vast, formed of this physical objective world and much more. Since there is a difference between the actual human reality and the current knowledge about the human reality, while by altering the knowledge itself about the human reality, it splits it apart into the consensual human reality and the actual, normal, objective human reality.

While if you lack knowledge of all these, you end up exploited, as a human being, in a human world. Because you cannot study the human reality without comprehending those controlling the knowledge about the human reality and how they do so forcefully, consensually, or through stereotypes set in place since Aristotle and long before, since this is how they end up controlling you and the entire world.

Furthermore, you cannot conduct an accurate study of this real world, of this universe, if you do not understand yourself, if you do not understand your needs driving you to perform this study, if you do not understand your mind constituting your means of understanding the universe, and if you do not understand Life altogether spanning the universe, actively involved in its structure, shape, behavior and development.

Simultaneously, understanding this world is the key to understanding yourself, your life, and your meaning in life and in the world, closing this circle of knowledge. This is why we consider the most relevant circumstances behind the famous studies of this world, we find true ideas and how they influence the understanding of this world throughout time, we seek to understand how and why people accept consensual, scientific, and ideological models of this world and how this influences their life, interconnectivity, and development, while we discover systematically this entire world, in all possible details. Furthermore, we use this study of the human reality to test all significant knowledge and ideas including human reasoning, past civilizations, indoctrination, Einstein, astral planes of existence, ideologies, Renaissance, the Brotherhood, ideological control, ages of Earth, cosmology, cosmogony, social control, mind control, Giordano Bruno, consensual interconnectivity, relativity, human origins, human development, Copernicus, the Consensual Matrix, the Big Bang, dreams, ancient wars, stereotypes, Galileo Galilei, conscious reasoning, Schrodinger, his cat, creationism, alternate realities, and much more, the entire human reality.

This book studies systematically the human reality, focusing on accurate truth while discarding beliefs and errors of reasoning, correlating with all relevant knowledge form physics, religion, spirituality, society, education, history, psychology, and more. If you want to learn more about everything surrounding you and everything that you really are, this book is for you.

Everything relates to your inner, natural needs for learning and development, exactly as you fulfill them right now while reading this book. Because through these genuine developmental human needs, you are capable to gather the accurate knowledge necessary to develop yourself and tend to the intelligent human environment, making this world a better place, an equal, just, prosperous, compatible place to live in, for you and for everybody else. While many times, it is a strong, inner desire to develop and make the world a better

The Human Reality

place coming through all your cognitive abilities, including your entire highconscious mind, which is your higher self.

Consensual means by agreement, while you must have two sides or more in any agreement in order to be able to instate anything consensually, as it always remains abstract, unnatural, unreal, fiat, and therefore not there, not part of Life, and not part of the real, objective world. More precisely, the real cannot be consensual, which means that reality cannot be fiat or consensual, since it is actually real, regardless of how many agreements, oaths, statements, signatures, permits, and certificates you involve to make it anything else consensually. Life certainly disregards the consensual entirely, since it is not real, neither natural, nor alive, yet the consensual attaches itself to this world and to its people by law, through specific laws, agreements, and beliefs that it makes itself, in a major conflict of interests. This is very important to study, since it actually contorts the world while taking over.

Because when I divide the human reality in the consensual human reality and the actual, natural, normal, living human reality, it is not exactly an allegory, but it is the actual case. The human reality is always divided into the consensual reality and the natural reality ever since the Romans, the Mesopotamians, the Sumerians, and long before, and it is important to be able to identify the human reality exactly as it is, because science, media, education, and the Brotherhood will never tell you the truth of your own reality and of the human reality in general, because they are not allowed, and because they do not know it themselves, regardless of what they might assume.

Yet how can the consensual even influence the human reality in any manner, if it is only abstract, fiat, or not there? It does not, yet the people do, as they remain constrained consensually to change and contort the world to everything else, against their actual normal needs and behavior. Now this is the human reality, contorted, into something else, while you cannot even identify anything different here, since nobody else does. Why should you? How can the world ever be consensual? It makes no sense, no?

Let us take society as a reference, since it is a direct part of the human reality, it is real, and it is there, with all the people always around. Yet is society actually real? No. The living human beings are real, along with their continuous interconnectivity, and so is your family at home, while the current society itself is not real, but only consensual. Because just the way you lack discrimination and exploitation at home in the family, you should lack them in the real human society. Furthermore, you tend to blame politicians for all discrimination and injustice in the world, while these are only marionettes in society, consensual puppets. Since just as stated above, once you fail understanding the human reality, you end up exploited as a living human being in a human world, not exactly because there are corrupt politicians in the world, but because you have the entire consensual in the world, replacing systematically the real and the alive, while all consensual lies replace the truth.

Discrimination leads to exploitation and even to eradication, and it is the case because the current human society is divided consensually into three separate social classes. While stereotypically, you always accept society divided into social classes, because why not? Yet social division means social discrimination, leading to social exploitation, and even to social eradication, which is what you have currently in the world. While this is not actually part of the normal human reality, but of the consensual human reality. More precisely, you are constrained consensually to exploit or to be exploited, throughout an unjust world, not exactly by force as you might assume, but consensually, through your own agreement. Do not underestimate the consensual, since it is not exactly as fiat and as abstract as it states, but it is an entire Consensual Matrix spanning most of the wider world, affecting you continuously from birth to death, life after life.

Who are you exactly? You are certainly the natural, living, intelligent human being, yet notice how the multitude of your documents never represent you as a natural living human being, but only as an artificial being, fiat being, strawman, or consensual corporation. This makes all the difference in the world, because through the multitude of your documents, licenses, permits, certificates, and ID cards, you are part of the Consensual Matrix continuously, abiding only to its rules. While you are the natural living human being and part of Life and of the real world only through your obvious human nature, but this is always ignored in the current consensual society.

Can the human reality itself actually be consensual, fiat, and not there? Yes, certainly, if you ever allow it to be in this manner, because you can always agree on anything with anyone you want. You might still ignore all these if you are from the Masses, because if you are from the Brotherhood or from the Elite, you already know the truth, since this is basic knowledge in the upper society, being used against you, as you are kept under control, consensually. This is why now we do not have only a natural real world to model here in this book, but we have adjacent artificial and consensual worlds that we call reality by agreement. As you can already notice, the artificial and consensual worlds we call reality are more important for humanity and for the entire Consensual Matrix than the actual natural real world, since there is where they live, exist, and interact, in the Consensual Matrix.

Yet since ignorance and suppression of knowledge harm and even kill, now through ignorance alone, entire social layers are forced to hurt and oppress those below, consensually. They do so in order to keep the world as it is currently, consensual. People in power work overtime to maintain this consensus, this official model of the human reality unchanged, the way it is currently, the way it has always been, because once they step out of this enforced, consensual model of the world, the world starts developing and it remains developed for longer periods, or it is contorted and enchained once again by tyrants throughout future revolutions.

Why exactly is the world as it is today? There is a reason for everything in the world when you ask science, since it is the task and responsibility of science to offer all knowledge about the world, about this entire reality. Therefore, the truth is always part of the scientific consensus, but who exactly controls the scientific consensus? Consensus means agreement, consensual agreement. Understand the scientific consensus first, and you are able to understand what they hide, how, and for what purpose. Furthermore, science is supposed to offer all knowledge that we seek, everything related to the human reality, while it never does, it never offers the actual accurate truth, but only a consensual version of truth, which is never the same, because truth is always accurate and therefore unique, and it can never be consensual. Yet consensual it is, and it is even officially accepted in the world, by the entire scientific consensus forming the current mainstream science.

What is the difference between the consensual and the accurate? What is the difference between the living and the consensual? Let us see.

2 THE REAL AND THE CONSENSUAL HUMAN REALITY

Would you like to know everything about the human reality? Just study the current science and you learn everything, about the origins of this world, its metrics, matter, radiation, waves, stars, planets, gravitation, electricity, roads, mountains, water, fish, buildings, cars, and houses. Yet is this accurate knowledge? Some is and some is not, since you cannot know everything anyway, through human limitations. Yet was the big bang real, taking place right then, in the very beginning of the world, while forming this entire world? Yes, certainly, since all scientists of the world state so, in an entire scientific consensus, which means agreement, stating in this manner the beginning of the consensual reality. Yet we want to know the accurate knowledge, the exact origin of this entire world, because it defines in part the actual human meaning in life and in the world. Was the big bang real, or the world was actually created? The current scientific consensus provides only the big bang theory and nothing else, but only until they change their mind, to agree on something else, on a steady state universe more probably. Since again, we have the consensual interfering with the real itself, with the real accurate knowledge, affecting everything in this manner, even the human knowledge.

Who takes part in this scientific consensus, this globally spread scientific agreement? All scientists do, the entire academia, and everybody from public research centers, consisting everything scientific in the open, unclassified. How can you control the entire amount of scientific information generated globally through hundreds of millions of scientists, professors, teachers, and technicians? You cannot, but the Elite can, through the current hierarchic Brotherhood. This is the case in the West, while the tyrants of the East control the rest of the world, all done consensually, through your own agreement.

Who is bad and who is good? Is the Brotherhood bad? People are people, and cannot be divided into the good and the bad. Just try to divide the deer from the woods into the good deer and the bad deer, the fish into the good fish and the evil fish, or the birds, cats, and horses, since they are neither good nor bad, but only normal living beings.

Yet you can always divide living beings into the developed and the underdeveloped. The first developmental level relates to everything consensual, and it includes all servitude, hierarchies, ideologies, jurisdictions, and politics. The second developmental level includes everything animal, physiological, and intuitive. While the third developmental level is intelligent, while I refer to it as the intelligent human level. This was supposed to be the actual human developmental level, but as you look around, you see people living life underdeveloped, mostly drugged at the zero addicted level or in tyranny or under servitude at the first consensual level, but also through raw animal instincts at the second intuitive level.

Yet while living your life under servitude at the first consensual ideological level, you are told everything meant to keep you under servitude, even that your specific ideology and jurisdiction are the best in the world and therefore you are always good within them, with the rest of the world not too good or even bad, depending on ideology. Since everything depends on the specific ideology in use, because all ideologies

The Human Reality

assure you that you are good in the world for as long as you follow them, with the bad people following everything else. This is the case with the Brotherhood, the Masses, and the Elite, since ideologies are in all social classes, as they count in hundreds of thousands, mostly fighting and contradicting each other.

Yet people are people, neither good nor bad, while only ideologies, laws, and jurisdiction can cause disagreements, discrimination, competition, and even exploitation and eradication among people. This always has a purpose, since ideologies are always controlled. Because it is a stereotype to assume that the world is divided exactly into good people and bad people, these put up a great fight until the end of the movie, order, or regime, the good people win guaranteed, this solves all problems, and now everybody can return happily to work. Yet this is not actually the case, because this world is more diverse even than the multitude of stereotypes and ideologies spanning this world, and you need to understand everything in the world and everyone involved before you can understand the entire world.

The current Brotherhood is neither good nor bad, but it is just a social class, the middle class, used by the Elite in every manner. While it is similar with the Masses, since everybody follows the same fulfillment throughout life. It only depends on who uses society and for what purpose, since as already stated, people are more or less developed in the world, while through their development alone, people remain more or less determined to help or to harm the world.

Furthermore, it is possible that you are in the Brotherhood yourself, in the Lower Brotherhood more likely, since there are not too many Masses left in the world. While what you are never told is that the Lower Brotherhood is the new Masses of the world, since there are not too many Masses left. This is done deliberately, only for you to be able to control the Masses, directly and rigidly, as this social control comes from the Brotherhood.

The Brotherhood is made of regular people, those who had

decided to gather in larger groups, legal or illegal, in order to help each other in every manner while competing in the world as brothers, as they even call themselves Brothers. While the Masses compete in the world individually or in small families and small circles of friends, standing no chance against the Brotherhood, and therefore always remaining the bottom social class. Yet as you study the world, the Masses are already extinct, either through eradication or through transfer to the Lower Brotherhood, while the Lower Brotherhood has to pick up the load of the world, becoming the Masses of the world.

The Brotherhood is the middle social class, right above the Masses, it is highly discrete, and it has members everywhere, in research facilities, educational institutions, industry, medical research, finance, and politics. Yet this is the case with all ideologies and even with all minorities, since these can be everywhere, and their members always tend to each other, wherever they are. Because this is the difference between competing in the world individually or in large groups.

What exactly makes the current hierarchic consensual Brotherhood so successful? Discrimination. Because what you can never see and what you are never told is that by helping only those within the Brotherhood, you discriminate those outside the Brotherhood, mostly the Masses. While the Elite plays this entirely rigged social game on its own behalf, above Masses and Brotherhood alike.

Yet you already have in you the necessary needs to develop yourself and the world, and to assure equality, justice, and abundance in the world. Many times, this is why you join the Brotherhood, in order to be able to develop and to help the world, according to the Brotherhood ideology. This is the case with benefic brotherhoods, because criminal brotherhoods have a different agenda. Good, bad, legal, and illegal Brotherhoods always interconnect perfectly, to form the current consensual Brotherhood, which is the second social class of the current consensual society, while discriminating and therefore exploiting the rest of the world, with all Brothers constrained to comply, within very strong social hierarchies

The Human Reality

spanning the Brotherhood and the entire society. This is the actual facet of the human reality, and this is what you always have to obey.

You always have to take the oath in order to be part of the Brotherhood, and through this oath, you even give permission to the rest of the Brotherhood to harm you and even kill you if you happen to diverge in any manner from the Brotherhood. Therefore, you are bound to the Brotherhood in everything that it demands from you, and this assures to the Brotherhood all success and all cooperation that it needs, which is unconditional obedience.

Who exactly are these? You are very familiar with the criminal Brotherhood, consisting of all mobs and criminal cartels of the world. You are also familiar with the knights, the mason, and everyone alike, forming the current legal Brotherhood. While if you are from the East, you are certainly familiar with the communist Brotherhood behaving similarly throughout the dictatorships of the East. With some exceptions, yet it is the same everywhere else in the world, since the entire world is filled with tyrants, covert or in the open, yet always forming these legal and illegal Brotherhoods in the upper part of society controlling everything below while serving minutely an Elite above, which seems innocent continuously. Yet regardless of how they are, legal or illegal, covert or in the open, all branches of the Brotherhood remain compatible and help each other in every manner throughout the world, while exploiting and controlling tightly themselves and the Masses below, mostly consensually, but also forcefully.

The Brotherhood does not control you from a distance by drastically enforcing decisions or by infiltrating key institutions in a stealth, ingenious manner, as you see in the movies, because the Brotherhood is these institutions altogether, since everyone working there is a member of the Brotherhood and a follower of higher orders. With the few who are not becoming scapegoats soon, to take all the blame and to lose their jobs, and to work in restaurants and in superstores from then on.

Furthermore, most of these institutions had been created

and built by the Brotherhood to begin with, and now they belong to the Brotherhood altogether, along with everything else. Because when you are in the Brotherhood, it does not matter where you work, where you go to school, and even who your family members are, because when you are in the Brotherhood, your legion is for the Brotherhood alone, and you do exactly what the higher Brothers order you to do, unconditionally. Therefore, it does not matter if you lead a university, a battalion, a bank, corporation, charity organization, political party, nation, art studio, government, hospital wing, senate, kindergarten, or apartment complex, you are never part of these as it says in your job contract, you never act according to their specific requirements, but you do only what your upper Brothers order you to do, because you are always a Brother first. Therefore, do not blame politicians anymore, since these are in the Lower Brotherhood, always doing as told, and always happy to get some crumbs, while you work very hard to do as you are told in the Brotherhood, regardless if you harm everything and everyone at work, in the party, at school, at the factory, or in the entire nation. You do everything according to the Brotherhood, and you are always very proud. While if you are from the Masses and this is the first time that you learn of all these, it is certainly easier to understand the Brotherhood by studying the Brotherhood from the communist countries consisting of the all members of the communist party, or by studying the Brotherhood of the old Germany consisting of all the nazis, the gestapo, and the ss, only that it is the same in the West with the entire masonry, since it is the same Brotherhood spanning the world, one dark age after another, while they always cooperate and they always share the profit.

While you see in the movies exactly the opposite, with everybody portrayed as though they act on behalf of their institutions, places of employment, political parties, and social organizations, and as though everybody involved is part of the Masses. Because while the Masses are always in the open, whatever the Brotherhood does, it is always covert. The

The Human Reality

Brotherhood is even hidden from the news, from songs, from the movies, and from cartoons, while most of the world is in the Brotherhood, by the billions, and they are always hiding. More precisely, the entire activity performed by the Brotherhood in society is hidden, with the entire social class above the Brotherhood, the Elite, hidden altogether.

As you know, the current sociology depicts the human society as being formed only of Masses, divided in three social classes, according to income. Yet this is not the case, because there is an entire consensual society running in the background through Brotherhood orders. These consensual Brotherhood orders ruling the world are unconditional and you must always obey them, while these are the actual artificial, consensual needs and meanings interfering with the fulfillment of all natural needs and meanings in the world, while this is how the natural human reality is always harmed and constrained by the consensual human reality. It is done on behalf of the Consensual Matrix, with all the pride and abnegation coming from the part of the entire Brotherhood and even from the Masses below, who always serve.

It is amazing to find people in the Masses who ignore the actual human reality. If you are still green and just out of college, and if you have never heard of any of these before, if you assume that the world is as depicted in soap operas and in the evening news, and if you still carry this strong, unbearable urge to become the famous medical scientist that finds the vaccine to cure cancer and diabetes forever, or to become the physicist to invent the most efficient sausage machine, capable enough to feed the entire word, the actual sausage machine that makes food cheaper and abundant to everyone, eradicating famine in the world altogether, and if you can never find a job in research and you still work at the same restaurant ever since you were little, this is the case not only with you, but apparently, you can still find people like you in the Masses today. Very, very few, but they are still there, ignoring the entire masonry in the world. Because the human reality is significantly different than what you learn in school and see in the movies, while it is still different than what you are taught at the lodge if you are a Brother.

What else should they tell you at the lodge? About aliens and Ponzi schemes? No, but about the Brotherhood itself. Because the current consensual Brotherhood is a consensual social matrix in itself and not only an entire social class offering you the opportunity to conduct your entire consensual activity. You can conduct this in any manner you please, as long as you abide by the Brotherhood rules, since these consist the actual Brotherhood matrix. Everybody already knows this, while the upper Brothers will never let you know how far they are capable to use the entire Brotherhood social matrix on their behalf. In this manner, you will always serve them, since this is the only thing that you know how to do, for a small profit, your actual crumbs, and you are always happy. While ruining the world in the process, yet who exactly cares anymore, when care itself is burned at the lodge throughout rituals? No more care, and therefore no more remorse while exploiting and trashing the world.

Yet even this is not the real facet of the human reality, but only a very small detail. Because the Brotherhood is only consensual in nature, fiat, it is only a small part of the Consensual Matrix that spans countless of worlds and realities, while the entire Consensual Matrix is consensual and therefore not part of Life but it is only crumbs of Life, along with all consensual agreements and all fiat consequences that these bring.

Because the world around is more diverse, and Life has sent you in this world specifically with your strong determination and higher knowledge and ability to help, save, and improve this world. You actually have the necessary natural intrinsic needs and meanings to do so, just the way you have the specific intrinsic needs and meanings to tend to your family and all your loved ones. Yet the entire world that you are supposed to save gets in your way, stopping you from fulfilling your natural meaning and save it, because the world has its own consensual wish, mostly financial. While Life respects it,

The Human Reality

and focuses her attention on other natural matters, even by discarding you in the process. Let us see how everything takes place.

Science was supposed to research minutely this entire world, while science is full of hierarchic Brothers. These already have their own very specific agenda, and they simply research exactly what they are ordered from above to research, mostly universes of sixteen dimensions and entire big bangs that never happened. Coincidentally, they obtain exactly the same scientific knowledge and results as ordered from above, and you learn it in school, as everyone else, Masses and Brotherhood alike. This is how science remains integral part of the Consensual Matrix, while it was supposed to research a genuine, natural, living reality, all part of Life and not of the Consensual Matrix. This is how everybody abiding by the consensual knowledge of science remains part of the Consensual Matrix, either directly or implicitly, since the knowledge offered by science never helps you integrate in a natural, living, harmonious society, but only in an artificial, consensual, hierarchic society, also part of the Consensual Matrix. Coincidence.

Everything remains consistent and very steady, lasting for ages in a row, the actual dark ages. While as stated, it does not matter if it is the Brotherhood of the old Germany with all nazis included, or the inquisition and nobility of the old European aristocracy, the communist Brothers from the communist parties of Asia, or the criminal mobs of Italy and Orange County, because it is the same Brotherhood, legal and illegal, allied closely throughout the world, currently and throughout history, making possible all tyrants, all dynasties, all ideologies, and therefore all the dark ages, one after another after another indefinitely. As a reference, most of the Consensual Matrix consists of medieval dynastic tyrannical worlds, countless in number, made possible by similar legal and illegal Brotherhoods, in the cruelest manner.

You cannot divide the people of this world and of the entire wider world into the good and the bad, or into the legal and the criminal, because these are the people themselves, and this is how they interact. Therefore, when you study this world and the entire wider world, you must study everything in both their real and in their consensual form, for a clear, accurate, consistent understanding. While currently, the entire society hides most of the knowledge of this world for various purposed, and if you ever want to learn the truth, you must find it and you must elaborate it yourself.

At the academic level, scientific knowledge is controlled through a list of references, credibility levels, and strict funds. The funding issue is only an official reason to approve or reject research projects, and this is how irrelevant research projects are approved, with important ones being rejected and therefore having no funds.

Who is on top of this system? The Rockefellers are in the West, while there are other elite families above them deciding everything, with the dictators of the East doing the same in the rest of the world. However, currently, the dictators of the East take over the entire world, while taking the invisible kingdom down. The Rockefellers do not control only the dollar, the banks, and the oil industry in the Americas, but they control all war industry, all medicine, food industry, charitable organizations, all research centers and the entire education, not only through funds, but through well-defined lists of scientists, research projects, and controlled results. These lists of references count the most in the world of science, since these scientists are the ones forming the scientific consensus, and they are Brothers. Yet since all Brothers have their legion in the Brotherhood and never in the mainstream science, now this is the kind of scientific knowledge that you learn and teach in school, exactly as ordered in the Brotherhood. However, since you are also a Brother as a teacher, you are very proud to teach everything irrelevant, harming the world some more.

Yet there is more to consider, because in the West, there is a very large group of people competing in the current society as a large nation, an invisible nation, an invisible kingdom actually, and the Brotherhood itself stands no chance competing with it, falling in servitude to the invisible kingdom. This is the invisible kingdom, originating in Georgia from the Caucasian region and now spanning the world but mostly the West, controlling the world as it pleases from within the Brotherhood and probably from within the Elite, by controlling tightly the legal Brotherhood itself, along with the illegal Brotherhood. With the most famous names of the world in the invisible kingdom and therefore in the Brotherhood, Rockefellers included. While when you are in the invisible kingdom, your legion is with the invisible kingdom, not with the Brotherhood, nor with your school, office, political party, military, police, university, scientific research institution, city, nation, kindergarten, hospital, alphabet agency, or financial institution, but exactly with the invisible kingdom. While the rest of the Brotherhood stands no chance competing with the invisible kingdom, but has to serve, handing the West to the invisible kingdom.

Yet as you notice from the news, the dictators of the East fight directly the invisible kingdom, seeking to take the West for themselves, to rule the entire world. Yet since the invisible kingdom is worthless, it has no chance, letting the dictators of the East win.

This is actually the human reality, very discriminatory and very abusive, while you never learn this in school. Because the invisible kingdom owns science, education, academia, entertainment, media, finance, economy, and even history, food industry, and politics, and does not allow you to learn this in school. This is the case in the West, with the dictators of the East doing the same, because in an undeveloped world, people will always fight to their death only to be able to serve more, since this is the consensual in contrast with the real, or the undeveloped in contrast with the developed.

While in the meantime, as a scientist, you must go with the main stream of science and do as you are told. You must be on the pertinent lists of scientists if you want to keep your job, your family, and even your life, since this world is highly organized on its hidden side. If you are within the scientific consensus, do not expect to bring major contributions to science and save the world, regardless of how wonderful you are, since your research, as it follows the scientific consensus and the approved lists of references, will give you no pertinent scientific results, all being irrelevant and misleading. This is how this world has all its theories as the big bang theory along with universes of thirteen dimensions curled up upon themselves, while people die in many poor nations of shortages of all kind. Because there is nothing relevant in those lists of references to improve science, to improve this world, and to save lives in this world, since they provide only a consensual dead end, keeping you distracted and astray.

Scientists laugh at you when you try to work on something new and useful for the world, yet they count in tens of millions or more while you are only one, and you must abide and do the same. Then they laugh again some more when they kick you out of the world of science and academia, and even out of the world of education. Go back to work at the restaurant now, if you are still from the Masses.

This is the kind of discrimination taking place in the world, while people are made to assume that they are wonderful and highly helpful to each other just by being in the Brotherhood, in the invisible kingdom, or in the Elite.

Because this world is meant to remain as it is, and the people themselves are highly motivated to keep the world in this consensual manner, even through intrinsic needs. Would you like free of cost energy and abundant food for everybody in the world? Well, it seems that the people of this world do not want this, and you cannot go against the world while trying to save the world, because you end up going against the people that you try to save in the first place, so good luck to you and to the rest of the world. While the world remains as primitive as it is, inevitably, since it is only a consensual world.

While with science on a dead track, there are no more new research ideas allowing you to work on, since noting new fits the current references and scientific consensus needed to back you up, while you are made to believe that you are incapable to find them, being too stupid. If you persist and if you really want fame and recognition, you will have to work with very advanced quantum physics, researching useless little theoretical strings of energy and little universes of thirteen dimensions curled up around every point of space, which will never replace the gasoline engine in your car and will never feed the world, but they only steal the money of this world by the trillions of dollars, harming this world even more.

Do as you are told, follow the scientific consensus, get yourself on the lists of references, and you become what you always dream, a wealthy physicist, university professor and famous researcher, highly rated in the world of academia, with a few worthless theories and discoveries already associated to your name, with money in the bank, a larger house, a younger wife, and friends and relatives on high places, all members of the Brotherhood just the way you are. You have been spending public money throughout life yet you have never contributed positively to the world in any way, while you have already taught irrelevant and misleading information to thousands of university students who now follow on your footsteps, since they are also in the Brotherhood. Furthermore, you took the job from someone who could really discover something important for the world, who could have really saved the world, probably just the way you once wanted to do.

Do not worry, because you are not alone, since the entire academia is just as you, in the Brotherhood, with some or most of them in the invisible kingdom. Study the consequences, since the world stagnates for decades now, and it seems to start decaying. This brings down the Western Civilization, along with the entire world, for the Eastern Civilization and many of its dictators to take advantage, and prosper globally. Yet nobody cares, with all care burned at the lodge, even if it happened before in the West and East alike during all dark ages since history is full of these, and nobody cared. Why should anyone care, when the Brotherhood is similarly everywhere, throughout all dark ages? Because it does not matter if it is called masonry, communism, mafia, aristocracy, ruling class, isis, invisible kingdom, ss, or red shirts, since it is the same Brotherhood, with the same souls, the same profit, and the same tyrants consisting all these, with the Brotherhood cruelty and underdevelopment always in control.

At higher social levels, the current science is called exoteric science, the science for the Masses, the knowledge out in the open. On the other side, the accurate science, or the esoteric knowledge, is kept for the upper social class, for the enlightened ones, it is highly expensive to produce, it is paid by the Masses directly from the budget of each nation, and it is legally secret and therefore classified, out of reach for the Masses, while this is discrimination.

Note how, despite of how hard the current Brotherhood works to keep the Masses misled, astray, and diverted, they never have the chance themselves to enjoy the products of the real science, regardless if they work on accurate scientific research themselves, since the Elite above takes all results, all credit, and all benefits, using all meaningful knowledge for themselves, as they even consider themselves a separate, divine, highly advanced civilization, apart from humanity itself.

Therefore, we have just identified the bad people in the world causing all the problems, the Elite. Yet this is not exactly true, because the Elite is controlled and exploited from higher above in the Consensual Matrix. Most of the science and technology developed on Earth is banned and forbidden from beyond Earth and from the worlds above, because humanity is not allowed to develop to galactic levels, or probably not yet, as it is always told. The rest of the civilized life in the galaxy might or might not know this, they might or might not know about Earth, since they are lied to in a similar manner, all being part of the same Consensual Matrix spanning this world along with many realities above. Because civilizations as the human civilization raise and fall continuously throughout the wider world, everything is controlled tightly through the Consensual Matrix, while everybody is made to believe that the rise and fall of entire civilizations is only a natural process.

While it is not, since everything is part of the Consensual

The Human Reality

Matrix, and nothing is random within the Consensual Matrix. What the wider world believes is that these young civilizations are always too ignorant and too incapable to make it through, so they can never survive, because the galactic environment is too harsh for them, and consequently, they perish with each significant cataclysm, so too bad for them.

While specific forces controlling the galaxy ban all worlds from developing the necessary means to reach the necessary advanced level allowing them survival, subsistence, development, along with a decent lifestyle and a decent civilization. It is the same everywhere in the wider world, because the Consensual Matrix is everywhere and behaves similarly as it does on Earth.

Have we just found our wrongdoers at last, these bad forces spanning the galaxy as part of the Consensual Matrix, deciding continuously the death of entire young civilizations just as it happens in Star Wars? No, not exactly, because these specific consensual forces are embedded in the entire galaxy from the most advanced civilizations to the weakest ones, with the human civilization caught right in the middle. The human civilization carries these same dark forces within, being innate and intrinsic, since these dark forces are intrinsically part of all civilizations, and probably even part of Life herself, as part of the comprehensive inevitably failure, lack of knowledge, and lack of development defining the weaker part of the living world. While this entire weaker part of the living world allows entire unfortunate living structures as the Consensual Matrix to form and decide the faith of most of the wider world, with Earth included.

These are the yin and yang of the entire wider world, yet in an unfortunate manner, because Life cannot know what she does not know in order to allow her to remain successful and therefore harmonious and fulfilled continuously as yang. Therefore, even if yin and yang are together in Life in an unfortunate manner while even being a supreme characteristic of life, you should not treasure the yin just because it is part of Life, but you should seek only the yang, to the point where you should have only the yang, since this is actual living real meaning, harmony, and fulfillment.

This is why we must include both the yin and the yang in our study of the human reality since this is the case in the real world, with the entire consensual human reality part of the yin, which is the weak, the undeveloped, and the unsuccessful part of the world, and with the actual living, natural, real human reality part of the yang.

Life does not treasure her lack of success, and neither should you. This is why Life invented death, in order to remove the unsuccessful, the meaningless, the old, the idle, the undeveloped, and the obsolete from the world. Development itself is a continuous intelligent movement from yin to yang, while development itself is also a supreme characteristic of Life and a law of this universe. The Consensual Matrix is not Life, but it is only an agreement standing outside Life and the real world, many times going against Life and the real world, and if you are not careful and if you remain consensual, undeveloped, and addicted, even deliberately, you are also against Life and the real world. Life can never cause the death of entire civilizations, as young and impotent as they can be, because Life invented death herself as a mean to develop, improve, and outperform herself, but not to kill her young, developing, successful self.

Currently, your cells die from within through a natural processes made possible by free radicals, meant to kill cellular components and even entire cells if they are compromised or if they are too old, after a specific number of divisions, if they are not successful, if they lack fulfillment, or if they remain idle, yet notice how even through free radicals, Life removes from herself only the dreadful, only the yin, but not the young nor the developing, and therefore you should never treasure the yin. It is the same with the entire organism, since it always dies through natural processes, mostly when it is incapable to fulfill its needs and meanings in the outside world. Entire species go extinct in this manner, with the best of the best to survive, as in the musical chairs scenario, because the environment is constant. It is similar in business and in finance, the same dreadful nonliving unharmonious musical chairs scenario, yet this is the consensual part of this world, constraining the entire human civilization in a dreadful, unreal, nonliving, and unintelligent manner, since this is the consensual.

Because as you notice, the problem is not exactly that the upper social classes plot against the lower ones. The problem is not the hundreds of millions of scientists working on irrelevant research projects while laughing at you and at your struggle to research something useful for the world as they should also do, but the problem is the laughter itself, the destructive attitude itself, and the consensual constraint itself, but not exactly the people, since everybody is involved in everything consensual. While if you ever go after the people to solve the problem, you end up harming an entire world, while this is not what you had in mind. This happens with the change of radical regimes throughout all nations during revolutions, since many people die in the process, while they were not even the problem, because there are always new harmful regimes to come, compromising even more people and getting them killed in vain.

The problem is certainly yin, not yang. However, as you study the current yin and yang closely in this world, you notice how the problem is the continuously enforced deliberate choice to have only yin in the world, only the consensual and the harmful, but not yang, because any harmony, prosperity, and abundance uplift and develop this world, causing everybody to stop serving in this world, while in this manner, making all tyrants impossible. While since this is not what the tyrants want, everybody remains undeveloped, meaningless, unsuccessful, nonliving, unintelligent, and unfulfilled, the entire yin, deliberately, exactly as demanded and expected from above.

While you cannot blame only the tyrants themselves, since they are only an effect, while you should blame your continuous drive to serve them while remaining undeveloped yourself, part of the yin of the world. because if you blame only the tyrants and the entire Brotherhood serving them closely, you end up replacing them with other tyrants and brotherhoods one dark age after another, exactly as you see throughout history, unsuccessfully. The problem is your servitude itself, since through your own servitude you make all tyrants and Brotherhoods possible even indefinitely, while you are always part of the yin.

This undeveloped attitude to serve tyrants alongside entire consensual Brotherhoods is artificial and even consensual, based on entire ideologies that determine and constrain everybody to serve. The old European aristocracy of the past two thousand years used their religious ideology as a platform to rule tyrannically, while keeping the West in an entire feudalistic dynastic dark age, and it seemed that it would never end. Later on, the invisible kingdom replacing them used science itself in form of a scientific consensus that is an actual scientific ideology to rule the world in a legal juridical manner, which is still the case in the West and even in the East. While the dictators of the East currently replacing the invisible kingdom through yet another world order will keep using their political ideology to rule the world in a similar tyrannical manner and through a similar hierarchic consensual Brotherhood, this time comprehensively, at a global scale, in a radical dynastic manner, exactly as Kim has in North Korea. While regardless of the world order and dark age, you always have tyrants, consensual hierarchic Brotherhoods, austerity, misery, loss, underdevelopment, and ideologies in a consensual conspiratorial manner in the world as integral part of the yin of the world yet always against Life and the real world, which is the actual consensual world that we study throughout this book, along with the real, the intelligent, and the alive, part of the real world.

The actual human reality is currently made of the consensual human reality and the real human reality, both consisting the actual human reality. The real versus the consensual, consisting the current human world. The consensual human reality versus the real human reality, while if you want to study accurately the entire human reality, you must study them both, the consensual and the real, but not only the real, nor only the consensual.

What is more harmful in the world, the current consensual Brotherhood, all these ideologies driving the world from one dark age to another, or this entire ritual of burning care at the lodge? Regardless of the question, the problem is always you, as you focus your time and effort on all dreadful consensual matters, since they only keep you dreadful, diverted, and undeveloped, while you were supposed to do something meaningful for yourself, for your family, and for the entire world, as developing yourself and the world, or making everything better and more meaningful and more harmonious, and entire human world.

Because even if the current human reality is formed of the consensual human reality and the real human reality, it is the case only through lack of success. Which means that you should never have the consensual in the world, because only the consensual creates all tyranny and all dark ages. While in contrast, Life herself keeps you only in golden human ages, only harmonious, only meaningful, and only fulfilled. While all tyrants, all consensual Brotherhoods, and all ideologies might state that there must always be dread and destruction in the world for any possible reason, as the Phoenix bird that must always burn or as the entire world that must always end, since tyrants will always say anything, yet this should never be the case in a real, living, intelligent, harmonious, meaningful, fulfilling human world, in the actual real human reality, not in the consensual human reality.

Notice how, since you precede tyrants on all lines of causality because you always make them possible through your own servitude, you are always the cause and the problem and you are always to blame, because you are the cause, not the tyrants themselves. The tyrants are even busy with their opulent life and they had never heard of you, so you can never blame them. Yet if you ignore all these and if you blame the tyrants themselves, you tend to replace them one revolution after another always unsuccessfully, because as stated, the problem is not the tyrants themselves, since they are not the main cause, but they are only an effect, as you make them possible yourself, since you are the main cause.

Others blame the ideologies in use throughout entire dark ages, while all ideologies are fiat, consensual, not there, they are the windmills of Cervantes and they are never there, because ideologies themselves are not the main cause, but they are only used by tyrants directly to rule, even in cruelty.

Even all consensual Brotherhoods are not the problem, regardless of how they are called, even if they inflict all the misery and all pain in an entire world, since the consensus itself forming them is an actual cause, not the Brotherhoods themselves. While in general, if your Brotherhood is hierarchic and or ideological, it is consensual, and expect the worse from yourself and from the entire Brotherhood, regardless of what it might state.

As a reference, degrees, rituals, entire ideologies, and all oaths mean hierarchy, discrimination, exploitation, and extermination, since this is the consensual, with all possible tyrants filling up the entire hierarchy, and with a major tyrant right on top, as a dictator, since this is the consensual, defining continuously the consensual human reality, which is main part of the yin of the human world.

While the consensual itself has other causes making it possible, as continuous underdevelopment and continuous ignorance, while as you study Life closely, you notice how she removes continuously the undeveloped, the old, and the unsuccessful from life and form the world, while always seeking only the yang.

Consensus means agreement, while it is always better to live life in agreement than in disagreement. Yet at a closer study, you notice how people always form agreements against life and against the real world, because people must always agree on everything that is not the case in life and in the real world, since if they were already the case in life and in the real world, they did not have to agree on them. Declarations of war are dreadful human agreements, along with exploitation itself, since as you study exploitation closely, you notice how exploitation is possible only through major agreements uniting entire mobs legally or illegally to harm, rob, and exploit the world, while ruining the world.

However, if you are more developed, you never form agreements against life, humans, humanity, and the real world, while you cannot even be exploited by all those forming agreements against you, because through the human intelligence, you always escape them. They can still eradicate you eventually, while in this manner, the undeveloped people manage to remove the developed people from the world, while making the entire world undeveloped, consensual, tyrannical, and exploitable, one dark age after another.

There are many techniques to decay the world and to keep it undeveloped, and they are used continuously by all tyrants, under all circumstances. You can find the protocols of the elders of zion over the internet to see some techniques used by the invisible kingdom to rule the West, yet there are invisible techniques always in use, and you must also consider them.

Protocols are also agreements, integral part of the consensual human reality, while affecting humanity continuously in a dreadful manner, to the point of extermination. One invisible protocol is disconnection itself. Humans are kept disconnected in small families or individually, made to live life in small houses and apartments, while through the harmonious intelligent human nature, humans are supposed to live life in common, in an intelligent human manner, everywhere in the world, while making possible the entire real living intelligent human reality, not the consensual human reality that you currently have.

Yet you never have the tyrant himself coming to your door every morning to inform you that you must remain disconnected, but there is an entire invisible ideology spanning the world determining you stereotypically to remain disconnected, called egoism.

It is this attitude towards individual gratification at the loss of the entire world that renders civilizations extinct, and not the fact that they are too young and therefore too ignorant. Because all civilizations were young and ignorant once, while now they suffer of civilized amnesia. These are the dark forces, these specific selfish, hidden attitudes and entire ideologies as sentientism, capitalism, communism, totalitarianism, and egoism, present deliberately at the level of each individual, of each family, of each nation, and even at the level of each social class. With the current consensual Brotherhood being only an example, while everything fuels the Consensual Matrix, since the Consensual Matrix keeps them instated on Earth, only to keep the Earth undeveloped, in servitude, and highly profitable.

Yet the Brotherhood is lied to just as the Masses are, while they are kept astray and diverted even from within the Brotherhood, just as it happens with the Masses. These are the dark forces, those keeping the Brotherhood astray, diverted from their true meaning in life and in the world, which is to do good deeds in the world, since this has always been their meaning, the actual human meaning.

What meanings and who are those keeping the mighty Brotherhood astray, right from within? The Brotherhood itself never identifies these inner social forces keeping it astray, while the Masses are so confused of what goes on everywhere, that they are ready to engage and fight anyone and anything for the right cause, including themselves. This happens during riots and revolutions, and people get hurt, and people die, while the Masses take all the blame. Ignorance, making all tyrants possible.

Yet you can see those controlling the Brotherhood from within, because they have been the ones tempering with the true meaning of the Brotherhood this entire time, rendering them astray and unfulfilled. This is enough to render an entire civilization astray, diverted, idle, and on its way to extinction, marching merrily to its common grave, because the Brotherhood were supposed to be the most developed people of the world, together taking care of the world, as they used to do until centuries ago. Yet not anymore, and never again, since this is the last world, never again.

As they say, who controls the money controls the world, but there is more to consider, since who controls the Brotherhood controls the world. This is the case currently and for the past few centuries, while it happened before, repeatedly, throughout all dark ages of Earth.

This destroys the Brotherhood, and implicitly, it destroys the Masses and the entire world, including the Elite. While the entire show is controlled by other civilizations through the Consensual Matrix, throughout majestic hierarchies of civilizations and galactic influence spanning the galaxy, the universe, and higher realities above, the entire Consensual Matrix. All these happening while everybody is made to believe that democracy and equality are well implemented everywhere in the wider world, while everything is a lie, a diversion affecting the entire civilized life.

It might seem irrelevant to perform any kind of study under these circumstances, for lack of ability and authority to change anything, yet it is always worth studying. It is worth studying all social actors involved, along with all attitudes meant to render them determined to act in a selfish, destructive manner. While it is not hard to model everything, since we have all the necessary knowledge from this entire book series "Human." Furthermore, unless we manage to distinguish and model these social forces at play, this specific world that they create, along with their behavior and consequences in the world, we are successful to create an accurate model for this real world and of the entire wider world, since everything is connected, everything is One.

Is this good or bad, having a consensual, artificial world? All lies are bad, and not only the scientific ones, even while claiming that they are done for a higher purpose, to protect the people. Some say that the world is kept in ignorance in order for the people not to panic when they learn the truth, while others say that this is the only way to keep the upper classes in power. Some say that aliens control the top social layers of the world, and they threaten us with extinction if we do not do what they say. While others say that the Elite went mad, or lost control of the world, or were even exterminated, and now no one knows what is going on in the world anymore, with everyone in the Brotherhood fulfilling themselves continuously. However, look around, to see that the world is in a very stable state, and it had always been in this manner. More importantly, our civilization has always used similar modes of existence, even when those on top of society were replaced repeatedly, one world order after another.

Study society closely, to see how there is a consistency in all methods of control in use across time, across nations, and across social domains, a majestic, very elaborate plan. There are similar consensual models not only for science, but for society itself, and for the human condition, thoughts, lifestyle, needs, purpose, expectations, ideals, conduct, and morals. Someone up there really does a good job, and had been doing so for millennia and more.

The world is always as it is currently, and nothing will ever change. The Elite always wants the Masses in homeostasis, in a stable, controlled state, and they keep the Masses stable by all means, through induced ignorance, subliminally induced obsessions, fears, discomfort, phobias, abundant drugs, poisonous food additives, medication, famine, shortages, crisis, and oppression. Yet the Elite always stays hidden, while maintaining its consistency as though it is either very strong, or it is strongly controlled itself from further above, from beyond, by very powerful forces, by perfectly capable entities.

This is the Consensual Matrix, it is the one controlling all knowledge, development, meanings, and achievements in the world, among everything else. It is this structure of high social power distorting the model of the universe form the human knowledge as they please, with the main purpose of constraining everyone to their agenda.

The Consensual Matrix is only an extraordinary social tool spanning most of the wider world, meant to help all intelligent life interconnect in the most efficient and just manner throughout the wider world. It is only the contorted manner in which the Consensual Matrix is used harming the world, since as all tools, you use it in any manner you want, but mostly, you use it in any manner you can. However, since all underdeveloped people seek to contort all laws and rules of the Consensual Matrix on their behalf and in the detriment of the others, now this is the world that they create, dreadful and consensual. Yet since you cannot stop the Consensual Matrix anymore because it had been instated through very high powers, now this is a problem.

It is easy to detect inconsistencies in the official consensual model of the human reality in order to detect the lies, distinguishing in this manner why they are used and what they hide, while this allows us to create a true model of the world. While since our model contains social, cognitive, and developmental structures, it forms in this manner the true model of the human reality.

If the Masses and the Brotherhood ever learn the truth, they still cannot develop, they cannot do anything to change the world, since the Elite controls them tightly even at a personal level. However, there is something that people can always do to change the world. They can refuse drugs, entertainment, and ignorance, as they learn and develop continuously, even all the way to the intelligent human level. More people today have the determination to refrain from harming themselves, others, and the world.

What is the use of protesting against oilrigs, against oil pipes and oil tankers, when the Masses themselves operate these machines anyway? It is as Masses protesting against Masses and Masses fighting Masses, and it happens often. The Elite never work in the oilfields, and they never even go in the oilfields, since it is dirty and it smells bad. Yet the Elite never work anywhere, because the Masses do. The Masses make the world exactly the way it is, so how can you protest against the Masses, when only the Masses protest in the world?

What can people do to make a change? People who develop, reason, and understand, change jobs and move to work in places where they do not harm anything. Some people ride bikes and forget about cars, others plant gardens, some educate themselves with true knowledge, and then they educate others, towards an entire harmonious intelligent human world.

As an example, a few decades ago, when the Queen went to visit Quebec, which is a Canadian province, when the Quebecers came to welcome the royal precession in Quebec City, they simply did nothing but only turned their backs to the Queen, to look the other way. Passivity and lack of cooperation with those exploiting humanity can make a great difference in the world, it happened repeatedly throughout history, yet you never learn about it in school, and you do not notice it unless you learn, develop, and reason independently. What you can learn instead in this consensual world is to state your intentions clearly, while these are yours anyway, to scream and curse during riots for no reason, and to fight the police in vain. Why exactly doing all these? Because the riots themselves are controlled to take place in this exact harmful manner. Because if they remained peaceful, they made a positive difference in the world. Because you do not have to do anything bad in the world in order to make the world a better place, despite of what the multitude of ideologies state. All that you have to do is not to cooperate with those harming the world, and not to recognize them as your authority when they harm the world or even when they are worthless in the world.

Yet you can do so only when you are capable to distinguish the good from the bad yourself, without beliefs, stereotypes, and ideologies. Because if you still rely on authorities to tell you what is good and what is bad in the world, then this is how you are controlled to do dreadful things in the world, by convincing you that it is actually good to do dreadful things in the world, and you should do more. Examples are many to give, since it always happens in the Brotherhood.

As another example, humanity has been used and enslaved many times by powerful entities from beyond. There is an English myth when the people of those old times simply crossed their arms on their chest and remained passive, refusing obedience to that entity, regardless of consequence, while she just left, she vanished, and she became instantly obsolete. Because she was in this world through those people, through their beliefs, and through their faith and veneration, preceding them on all lifelines of causality. Consensus is an entire consensual interconnectivity, and it takes an entire group of people to conduct it.

Because it always takes the master and the slave in perfect agreement to maintain slavery and any discriminatory relationship instated throughout any kind of consensus, and if you simply do not recognize that you are the slave, then the master has nothing against you, but has to go find himself another slave willing to swear obedience and willing to recognize that is a slave. It might seem awkward, but many people are always willing to become slaves and to remain enslaved, if their ideology only states that they must behave in this manner, since they always obey. Furthermore, they will harm you if you do not accept to be a slave alongside them, because ideologies are very strong while controlling your thinking and behavior. It is the same with the royalty, because when you turn your back and want nothing in common with them, what can they do? How can she consider herself your Queen in Quebec City, when you go about your business and you forget about her? It is the same with the multitude of false deities to have paraded this world, since what exactly is here for them to do when people refuse to acknowledge them? While it is similar with all authorities including extraordinary brotherhoods and extraordinary elites and enlightened personalities of the world, because as long as you have never heard of them, as long as you consider them simple living beings, how can they ever admit that they have any authority over you?

This is the consensual world, since it is only through your own agreement that you remain determined to serve, while it is also through your master's agreement that you are accepted to serve. This is the difference between exploitation and cooperation, because at the first servitude level, people are always eager to serve or be served by others, while they find nothing abnormal in this entire exploitive social interaction. While at the intelligent human level, people always enter in harmonious relationships with others, helping each other as genuine brothers and sisters.

This is the difference between slavery and freedom, yet not too many people are interested in identifying it, since the capitalist ideology is strong currently, ruling the entire world. While the opposite of capitalism is never communism, since communism is just another ideology, while in contrast, communal living is genuinely natural, part of the intelligent human reality and not of the consensual human reality. Communal living is similar to the family living that you have at home, only at the size of the entire city and the entire nation, if this can ever make any sense at any undeveloped level. While it does make sense at the intelligent human developmental level, since only at the intelligent human level you can distinguish between the consensual human reality and the genuine, natural, living, intelligent, harmonious human reality.

As a reference, only a few decades ago, it made sense that killing elephants in Africa for fun and for virtue was normal, while it is different currently, after only a few decades, because the world has developed so much in the meantime.

As we notice throughout the book, the difference in the level of development is so distinct currently, that it draws a solid line between the consensual world and the genuine, natural, living world, with the people developed at the first servitude level on one side, and with those developed at the third intelligent human level on the other. You find this obvious distinction everywhere around, in people's behavior, attitude, thinking, lifestyle, desires, and even in the movies, at home, and in your purse, because not only the world and the human society are divided into the living and the consensual, but the entire human reality is, including the human needs and meanings, along with the human fulfillment, attitudes, expectations, and desires, and along with the humans themselves.

The Human Reality

Because there are always two of you in the world, the natural intelligent living human being, and the consensual corporation. Your consensual corporation does not exactly represent you in the Consensual Matrix, but it is you altogether in the Consensual Matrix while disregarding your natural intelligent living human being. While it is enough to be developed at the first servitude level to agree with this cheap social scheme directly or implicitly, since this is how you always serve, and this is how you always live your life, consensually, regardless if you identify it or not.

Notice how strongly the Consensual Matrix persists to keep you consensual or fiat throughout life, coincidentally just as much as those old deities persisted to keep people venerating them in the past, just as much as royalties persist to have you recognizing and following them, and just as much as all masters persist to have you serve them as a slave.

From a third level intelligent perspective, you notice how you must always be a tyrant in order to accept anyone to serve you, because their own continuous servitude to you ruins their meaning, fulfillment, and entire life, while ruining the entire world, only for you to have someone to button your shirt, feed you with a spoon, and comb your hair. At the intelligent human level, this entire idea of engaging into consensual agreements of slavery and servitude remains abnormal and therefore unnatural. You might not see it yet, but as stated, it is as repugnant as hunting large animals in the wilderness for pleasure alone, or serving a human being that considers itself superior to you, or owning a slave yourself to make it clean up the house and do the dishes, as people did in the past, and as many people still do today, because this is the consensual, and you must always identify it if you want to live your life at the intelligent human level, as you always should.

While from lower developmental perspectives, all exploitation seems normal, while you will always maintain it for profit, pride, virtue, and social acceptance. While many ideologies determine you to maintain exploitation in the world, to serve or to be served, and so you do. While many times, you get in trouble if you do not conform.

As another reference, if you are from the Americas and you go to visit Europe, you might be amazed by the beautiful castles and palaces that you find there, always admiring the Europeans for their outstanding civilization. However, when you go to France to visit these amazing palaces and castles, the locals surprise you with their repugnance for their own beautiful castles and palaces, since slaves and servants had built and maintained these at great loss and great misery, while making all tyrants possible throughout the entire European dark age that lasted for two thousand years.

The castles and palaces of France are the symbol of human slavery, human servitude, and human exploitation, while many France people are ashamed of them, since the French themselves tend to be more developed. You have to be developed at the third intelligent human level in order to distinguish exploitation and the entire harm caused by exploitation, otherwise you will dream for the rest of your life to be a prince or a princess yourself as you see in all Disney movies, to live in a palace and to rule an entire world, as all tyrants do, with Kim included. While it is unacceptable for the intelligent human beings to be represented by repugnant castles and palaces as a nation, the very symbol of misery, exploitation, servitude, and oppression. Do you still want to be a prince or a princess? Do you still want to replace Putin or Kim?

You can notice these examples yourself if you exceed them in developmental level, while if do not, they make no sense to you along with this entire book and book series "Human," and you are bored while reading them or you are even frustrated, for insulting your beliefs and entire ideology. Because who wouldn't want to be a prince or a princes, to live an entire beautiful opulent life? It never bothers you that people toil at your feet by the millions, while you must always some or most of them in order to make the others abide, just study Kim, Putin, Lenin, Stalin, Mao, Khan, Noriega, Hussein and the rest of the dictators closely to see it yourself.

If you are a queen, emperor, empress, or king yourself, and if you are developed at the third intelligent human level, you never want slaves and servants in your life bowing to you, brushing your teeth, making your bed and cleaning your clothes, since it is inadmissible to have anyone serving you, mostly as slaves. Their servitude makes you feel uncomfortable at the third intelligent human level, and you let them live their own lives, independent from you.

While as you look around, you notice how all children want to be princes and princesses, since this is what they see in the movies, and it must be amazing. Yet how many people really want to have slaves, serfs, and servants in their life? Not because you cannot use someone to mop your floors, cook your food, work the fields, and drive you around, but because you destroy their lives in the process, since all they do is work for you the entire life, even if you feed them well. Since in this manner, they fail their own meaning and fulfillment in life and in the world, they fail Life themselves, and this is the inadmissible part at your intelligent human level. Furthermore, if your slaves, serfs, servants, and even employees are harmed or die accidentally while serving you, it is even worse as an intelligent living human being, so good luck to you throughout your depression, because it might never end.

Was it not repugnant as a child? Is it still not repugnant as an adult? Have you even considered these? Is it not amazing how entire nations can already see it, mostly in Europe, while no one from other nations and social classes do? Can you now see the difference between the real and the consensual, along with all consequences that they always bring? Can you still read Robinson Crusoe, or even Hemmingway, along with many highly prized authors? Does it take an entire model of the human reality divided into the consensual and the living in order to help you see the truth? Aren't these the questions that people should always ask?

How much longer are you determined to choose to be a consensual corporation instead of an intelligent living human being? Let us see. Next time when your car is pulled over and the police officer asks for your drivers' license, when he looks at it and asks if that is you, what exactly will you say? Will you answer that you are your driver's license, the consensual corporation stated there in uppercase letters, and hopefully you will be forgiven and allowed to go your way? Because as a consensual corporation, the police officer has absolute authority over you, and is allowed to do everything to you as necessary, regardless if you want it or not. This is what the officer does once you answer yes, because once you state directly or implicitly that you are a corporation, you enter in a consensual agreement with him that he has complete authority over you but only in his particular jurisdiction and district, while you must always comply.

Are you bold enough to point to yourself and say that you are actually the living human being? Will the police officer be able to know the meaning of your answer? Many do, since they are well trained, just search it on the Internet to see it for yourself. Because as a living human being, you have your human rights, allowing you to do everything in the world necessary to fulfill your needs. Nobody has jurisdiction over you, because jurisdictions are consensual and can are never compatible with you the living human being. As a living human being, you live in the real natural world, and not in artificial consensual jurisdictions and corporations as cities, courts, jurisdictions, and districts.

Will the officer really let you get away, if you only point at yourself and state that you are the living human being? Yes, immediately, if you are from the Brotherhood, since you know exactly the right signs to display, and you are free to go. Yet if you are from the Brotherhood, you place a specific sticker on your car, and they never stop you, avoiding the entire circumstance. Since all officers are in the Brotherhood, and they are more than pleased to let you go, since Brothers will always help Brothers, and it fulfills them. Brothers never have legal problems, unless ordered from above to stand down, have legal problems, and take the hit, mostly to cool down the Masses.

The Human Reality

What will it happen if you show knowledge of all these, and you are from the Masses? Will they still let you get away, or it is better to keep your eyes closed and do as you are told, as the Masses always do? Just search it on the Internet, to see what everybody does under these circumstances. However, at your intelligent human level, it will certainly feel abnormal to claim a consensual nature instead of who you actually are, an intelligent living human being.

What would happen if you say that you are an intelligent living human being, and not a corporation as stated in your driver's license? What if you tell them that your name is not spelled in uppercase letters but normally, you cannot be that name anyway since you are an intelligent living human being and you have your intelligent human rights, and no one should stop you from travelling and from fulfilling your natural human needs, or if they do, they interfere with your normal life, with your normal needs, with your normal human status, and with your normal human rights? They are deliberately attempting murder by stopping you from fulfilling your natural needs, because as a living human being, you die if you do not fulfill your normal human needs. Tell them this if you dare, to see what happens. Because it always takes two in each agreement, and if one does not agree or does not agree anymore, then the agreement is broken, and no law can force anyone in any manner. In all cases, as a living human being, you always have priority under all circumstances. You are the prime reference, making your statements always accurate under all circumstances, because you are the main reference as a living human being. Yet in general, they let you go the first time, and then they report you and they take you from there, targeting you and your entire family in any manner, by unleashing the entire Brotherhood on you, and you soon die.

It is important to understand how the Consensual Matrix will never harm you anymore as a living human being, since the Consensual Matrix is very just and never interacts with living human beings. However, the Consensual Matrix marks you as unwanted or as treacherous in the system, the multitude of its servants are summoned upon you in every manner, and will never let go of you until they exterminate you along with your entire genetic line, after making you suffer excessively to give you as an example, otherwise the Brothers do not obey anymore. However, it is possible that you are already marked for departure in the system even if you are from the Brotherhood and not only from the Masses, regardless if you are the one harassing or the one being harassed, since the Consensual Matrix always plays to win while you stand no chance. While many times you even know it, mostly in the Brotherhood.

This makes the difference between you the natural living human being and your brand or corporation, which is the consensual self in society, the consensual corporation that you actually are throughout the multitude of jurisdictions where you exist consensually, since jurisdictions span this world, while forming the Consensual Matrix here on Earth. While if you already start seeing correlations with major ideologies, it is because ideologies are of the first level, and they are based on consensual beliefs that are fiat, not there, only consensual. Ideologies are always part of the Consensual Matrix, meant always to obey the Consensual Matrix.

3 ACCURATE AND CONSENSUAL TRUTH, THE DUALITY OF THE HUMAN REALITY

Why creating models of the world altogether? Can't we just grab a cold beer from the fridge and watch our favorite show, the one with gorgeous celebrities talking about the length of their skirt and bodily parts? Isn't that better? Isn't this the human reality, measured in all details? Why not taking some pills in the meantime, while having a blast of a human reality? Why not ordering a large pizza and eat everything at once? You can do anything you wish, since society always accepts you, mostly if you do not learn about humans, life, reality, and the entire wider world. However, how exactly can you fulfill your real intelligent human needs, if you have no clue why humans are in this reality, and of what are their needs? How can you ever develop? Only ideologically? These are relevant questions when you live your life at higher developmental levels, and this kind of knowledge and reasoning make a significant difference.

Questions about your higher needs and higher behavior might seem funny if you happen to live your life at the zero level of addictions, at the first level of servitude, or at the second level of animals and animal needs, just as seen above.

Are you really fulfilled while living your life on lower

developmental levels? Because if you are fulfilled, then this world should always be a perfect place for everybody, and there would be no more suffering in the world. Yet is it really the case? No, not at all, because you are a human being, and it does not matter who you are and how you live your life, since you will always receive your real intelligent human needs. You are forced to fulfill them by your intelligences and you are punished intrinsically if you do not, only that by taking drugs and by engaging in lower level needs, you can cancel your real intelligent human needs along with their associated punishment, which works only temporarily and with dreadful side effects, while you also ruin the world. This is why drugs do not actually make you feel good, or they do so only temporarily, before they actually ruin your life and feelings. It is the same with servitude, since it deprives you of your own natural life. It is the same when you live your life at the animal level, because it makes you feel as an animal and not exactly as a human being. Because you can manage to remain a human being as long as you live your life as a normal human being, while identifying and fulfilling your intelligent human needs and meanings in life.

Can't you just watch "Discovery" instead, to learn about everything in the world so you do not have to model it yourself? It also has pictures, so it is fun. Yet you do so anyway, while the knowledge that you find there is always within the mainstream of science, presented at the elementary school level, it is not entirely accurate and consistent, it repeats itself, you already know it from school and it got boring long ago, while it insults your intelligence.

Did you know that you are already creating a model of the human reality in your mind right now as you read this book? You have been working on it since birth and you still consolidate it with everything that you learn, with all knowledge that you acquire, and with every experience that you have. This is your inner replica of the world, the subjective inner mind world where you live your life as an inner self, and I study it here and it other books of this series "Human," since all models are connected, forming one single larger model, "Human." This specific model of the human reality helps you define and consolidate your inner replica of the world, which is an inner reality, with a consistency and with an accuracy that you cannot obtain from the mainstream. You are bored in the mainstream because the mainstream knowledge is not accepted anymore by your cognitive system as accurate and your developmental intelligences punish you with boredom and depression, while always forcing you to search more in order to learn more, as much true intelligent accurate knowledge as possible. While you always learn everything through elaborated intelligent mental models maintaining consistency with the intelligent accurate knowledge that you learn, if you find it first, and if you know how to elaborate it in entire intelligent mental models.

In general, mental models need accurate intelligent knowledge and perfect understanding of the laws governing this world. Without these, the mental models that you create are flawed, and will never give you real, normal, credible results. You also have to test mental models for redundant cases, and make sure that they give you the pertinent data already known. This is how you fine-tune a model, to be certain that it works well for all future simulations, for every test, and under all circumstances.

While it is important that the mental model works similarly in identical circumstances, using similar laws and giving similar results. If not, you obtain aberrations, dualities, lies, and stereotypes, and then you have to invent more lies to confirm the flawed results endlessly. This is what happens with the big bang theory, and it has been going on in this manner for some time, eating up trillions of dollars, through all possible financial schemes.

How much does it cost to feed the world even endlessly? Because trillions of dollars can certainly feed the world. Yet the scientific Brotherhood researches the big bang instead, which is only a theory, nothing accurate. A better question is how much it costs to make the entire living free of cost, for everybody. Can this be more expensive than the current consensual life? No. Why still using currencies in the world, if not for dividing society financially into social layers and social classes, while enhancing discrimination, exploitation, and eradication? How exactly do you expect to be controlled without money, shortages, drugs, mainstream media, and erroneous, misleading knowledge?

Would anyone join the Brotherhood in a world free of currencies and dogma, in a world where everybody fulfills only natural needs and meanings? Yes, yet intelligent people always join intelligent harmonious egalitarian Brotherhoods, as these always form in an intelligent human world, meant to help and maintain harmony, meaning, fulfillment, intelligent knowledge, and prosperity in an entire intelligent human world, unlike the current hierarchic consensual Brotherhood that seeks discrimination and exploitation, not an equal world.

Because if you as a police officer help only the Brothers while you prosecute only the Masses, you have exactly what the old Germany had during Hitler, an entire dreadful discriminatory, exploiting, eradicating nation. It is the same currently in North Korea, because Kim does not discriminate, exploit, torture, and exterminate his people in large numbers, because Kim is always busy with his drugs, dancers, and entire opulent life, but his own people, the Brotherhood of North Korea discriminate, exploit, torture, and exterminate the Masses of North Korea. It was the same with the old European aristocracy and the entire European dark age, it was the same with all dynasties from the past and distant past from Europe and Asia, it is the same with the KGB, and the entire political party of Russia and China, and it is the same everywhere in the world past and present, covert or in the open, because every time you divide society into social classes and social layers you have hierarchy, discrimination, exploitation, and extermination, to the point where currently in the West, the Masses are almost extinct.

Since this is the dualism, between the egalitarian Brotherhood and the hierarchic Brotherhood, because only the

Brotherhood can make everything possible, either consensually hierarchically discriminatorily and exploitive, or in an intelligent egalitarian prosperous manner, but not both simultaneously, since this is the actual human duality at the undeveloped level. While it is enough to have people divided into distinct degrees in the Brotherhood, in order to have a consensual hierarchic Brotherhood. This is the dualism between the real living human beings, and the consensual corporations. Between the real living world, and the consensual world. While you always find dualism in any undeveloped unequal discriminatory world.

Yet can you actually understand this duality from the human society and from the entire human reality? Because all totalitarian regimes never state that they are oppressive, discriminant, and exploitive, even though everybody knows it well, but only that they are democratic, egalitarian, and for the people, which is never true, while creating and maintaining an entire human dualism that complicates everything. It is the same in the West both covertly and in the open, with all politicians stating how egalitarian they are while they always lie, which is the case in the entire masonry as it spans the West and most of the East, not only in politics. This duality is everywhere, standing directly on the law of relativity and on this entire real world, forcing us to mental model both the consensual human reality and the real human reality simultaneously, while doubling the effort. This would have never been the case if everybody were developed normally at the intelligent human level, according to the intelligent human nature, in life and in the world.

There are many dualities in the world, they result from erroneous, deliberate, or double reasoning, and most of them are considered erroneously the highest achievements of the human civilization. Yes, currently, social layers, social classes, and the entire current consensual hierarchic society is considered erroneously the pinnacle of the human civilization. We will identify all dualities along with the entire consensual throughout this book, since these interfere with our understanding of the human reality itself.

Take the law of conservation of momentum as an example, stated by Newton. That law had been used endlessly for any phenomenon, experiment, occurrence, incident, and accident, and it works perfectly. The laws of conservation of momentum and conservation of energy are the laws the most used in real mental models, along with the rest of the laws of classical physics. Apply these laws to any model within the universe, and they work fine, they can model anything you want, flawlessly. However, you must expand these classical formulas for relativistic circumstances, as very high speeds and very large masses, among other circumstances. Do you see the duality? You have to use two separate laws, two separate concepts for the similar event, while this should never be the case in our world, while it should never be allowed in science. You cannot have two separate models to study the same thing, while obtaining two separate results. Something is erroneous, since both models, classical physics and the current theory of relativity are insufficient or erroneous to model this world in all circumstances.

You can still state that the current theory of relativity is a continuation of classical mechanics for relativistic circumstances, since this is what the mainstream does. We can also use relativity instead of classical mechanics for any classical circumstance, only that we have to work harder, since for the few significant figures that classical mechanics always uses, it is not worth the effort to use the theory of relativity, because we obtain the same results by using directly classical physics. However, in the current consensual science, only a few circumstances are considered relativistic, as the distortion of space by high masses, along with the contraction of time under relativistic velocities, and nothing else. You are forced to use the current theory of relativity only for these cases, while for the rest, you have to manage on your own while using classical mechanics, or this is what scientists do, and they always encounter errors. Dualism. Because the current science is only consensual, constraining you to remain only under dual circumstances, erroneously, deliberately, so you can never know the truth.

While you cannot study the human reality erroneously, because all knowledge and all understandings of the human reality stand at the base of everything that humans know and understand in the real world, forming the entire conscious intelligent human mind, while affecting the entire intelligent human reasoning. More precisely, if you understand the human reality erroneously, it affects your entire human intelligence, while everything that you do in life and in the world remains erroneous. This is why nobody knows the actual intelligent human meaning in life and in the world, which is as playing tennis but you do not actually know what to do with the little ball in the field.

We encounter this duality here because both classical mechanics and the theory of relativity are empirical models, since they had been both created to explain only phenomena happening locally. Both models are not comprehensive, and they are not generated from natural laws of the universe, as the conservation of action or the conservation of energy density.

Does this duality, these separate empirical models affect our understanding of the universe? Yes, yet this duality, among many, confuses us the most and forms a fake model of the real world, a wrong understanding of the real world, leading us to ignorance even when we are highly capable. Let us consider another example.

Neutrons are not stable particles, and they decay every ten minutes or so, into a proton and an electron. Study this decay closely, to see how the conservation of mass, the conservation of momentum, and the conservation of energy do not work when you apply them to the decay of the neutron. More precisely, the mass of the neutron does not equal the sum of the masses of the electron and proton after decay. Remember, all models of the world should apply accurately to everything from the real world, regardless of how small or how big they are, regardless if they are colliding billiard balls, fragmenting asteroids in space, formation of entire stellar systems, or decaying elementary particles.

The failure of the laws of conservation of energy and momentum applied to the decay of the neutron had puzzled physicists a century ago when they had first noticed it. Only Pauli, a great particle physicist of that time, was able to solve this mystery, and became famous, but only consensually. While decaying, Pauli states, the neutron creates not only a proton and an electron, but another elementary particle, an antineutrino, to explain the extra energy, and to balance the equations. Behold the consensual neutrino. Do not try to find this antineutrino, Pauli states, because it is impossible to detect. What a genius, because his theory will always be true, since as he states, you are not even capable to prove it wrong, by definition, so don't even try.

Note that Pauli did not detect this neutrino, but Pauli only invented it consensually, since his few friends immediately backed him up. Pauli was not alone, since Fermi, Dirac, and Heisenberg formed a local scientific consensus immediately, and gladly supported Pauli, while Pauli supported all their theories in return, since this is how the scientific consensual works. Now you have to learn them in university, and what a bunch of consensual scientists, now part of the scientific consensus itself.

This is how they used Pauli's bogus elementary particle neutrino to create their own new theories, to become famous. Another duality was born right then: you can use the laws of conservation of mass, momentum, and energy anywhere and continuously in the real world without problems, yet only when you use these laws for particle decay, you have to invent new, very little elementary particles in the process, which should be impossible or very hard to detect by default, yet which you can employ to balance your equations as you please. Case closed.

Yet even through these dubious strategies, the equations of mass, momentum, and energy do not balance, making you change the mass of your bogus particles continuously, and even to add additional, newer, even smaller particles in order to patch up the first bogus ones. While the world loves you, for all the good that you do in the world with these little particles of yours costing trillions of dollars, because this is how the world remains irrelevant, wasteful, distracted, and astray. The real versus the consensual, resulting in the Nemesis of the entire human reality influencing your life continuously in this world.

Because it is very easy to identify all discrimination and exploitation from North Korea along with the entire bureaucratic contortion taking place through the entire masonry of the West, yet it might be difficult to notice how the same duality contorting the entire bureaucracy of the world also contorts all knowledge of the real world, simultaneously, through the same consensual agreements maintained unnecessarily among all the people of the world, in an entire harmful unnecessary human conspiracy spanning this world and the worlds above.

It is tedious to prove a lie true, since it takes more lies to do so, but then it is just as tedious to keep your lies instated year after year and scientific principle after scientific principle, since you have to keep on lying with newer and newer inaccurate theories and discoveries. This describes the current science, and this is why there is nothing invented and discovered to eradicate famine in the world, because the world is kept diverted and astray with this type of consensual nonsense, yet the entire world accepts conspiratorially the entire consensus of the world along with all dualities that the consensus always causes in the real world.

Elementary particles have different masses, resulting in a variety of aberrations when you use the classical conservation of energy, mass, and momentum. While every time when you cannot balance an equation, you invent an elementary particle to balance it for you. If you need more mass in an equation, you invent a normal elementary particle. While if you need less mass, you invent an antiparticle instead. This is what all consensual physicists do, while they always validate themselves through the Brotherhood, since the Brotherhood offers the perfect trustful consensual platform to make it happen. While as you notice, it does not matter if you validate Picasso as the best painter of the world, if you contort all financial laws in order to profit more, or if you contort all consensual theories of physics in order to invent other theories just as erroneous, because the current Brotherhood can make everything possible while profiting considerably, always channeling money from below to above in the hierarchy of power.

One bogus particle had to be invented for the decay of every elementary particle, and in time, these invented particles formed an entire bogus family of particles, now called neutral leptons. Neutrinos are divided into specific flavors: electron neutrinos, muon neutrinos, and tau neutrinos, depending on what nuclear reaction they need to balance, while you will always experience dualities of dualities of dualities in a consensual dual world.

Why are these new elementary particles never detected, since we should be bombarded right now with zillions of these invented particles coming mostly from the sun? The lie goes on while amplifying, stating that these particles do not interact with anything, escaping in this manner all particle detectors. They are not charged electrically so they do not leave a trace, and they have a very small mass so they are very hard to detect. Then how are these particles kept inside the nucleus if the nuclear force itself does not apply to them? The nuclear forces still apply to them, yet they do not apply to them, depending on various consensual circumstances.

How are they part of the neutron to begin with? More lies require even more lies. Behold another lie. There must be another force in the universe applying only to these bogus particles, and now they have discovered a new bogus force in the universe. There is another nuclear force applying only to neutrinos, something that you have never heard before, something that they call the weak force. While everything comes at a cost of trillions of dollars, and while another thousand children die in Africa today of starvation and another thousand children die tomorrow until they finish up all these little particles, to start focusing on food in the world for a change, yet do not hold your breath. Yet let us spend several trillion dollars more on this new extraordinary force that we have just invented, the weak nuclear force, and hopefully we can detect it, or not, since this is how the consensual works, while profiting considerably. While the entire invisible kingdom cannot stop laughing, since everything is faker hiding in plain sight.

If you want to understand modern physics, you have to understand its entire setting, otherwise it misleads you. Particle physics and modern physics in general appeared over a century ago, when people still rode horses on muddy roads, and it will affect us forever, since the scientific consensus will never cease to profit, since physics itself will never get back on the right track to research meaningful intelligent topics, as it always should. In this manner, you will always use combustion engines in your cars, you will always go to work all day long, and in general, the world as you know it will never change. Yet there is particle physics in the world costing trillions of dollars while studying everything irrelevant and unnecessary in life and in the world.

The four physicists stated above were very prestigious, top of the scientific consensus spanning the world, and did nothing for the world, but only took the place of those who could do something meaningful in the world. This is also how we ended up with four forces in the universe: electromagnetic, gravitational, strong, and weak. The strong force was already bogus, still being considered separately from the gravitational force, since the gravitational force has been badly understood from the beginning, being a stereotype passed to us from generation to generation ever since Archimedes, Galileo, Kepler, and Newton.

Yet this is not the real science, but only the consensual science, and it is used to eradicate the Masses directly through wars, medicine, and poverty, and implicitly by spending unnecessarily the common human wealth through irrelevant research, while it was supposed to be used for human survival, subsistence, development, and prosperity in life and in the world. While there is accurate physics researched in places as Area 51, remaining inaccessible to the Masses and the Brotherhood, regardless of what the Brotherhood is told, used only by the Elite, as they control the Brotherhood. With the Brotherhood working on all bogus projects throughout science while keeping the Masses astray and austere, and therefore weak, sick, hungry, and eradicated, while being very proud. This is the human reality, sad.

Why exactly suppressing the important accurate intelligent human knowledge? For military purposes, obviously. What military secrets exactly, when all wars in the world are staged? Can wars themselves be part of the deliberate continuous human austerity? Could it also be possible that politicians lie? Knowledge is hidden from the Masses and the Brotherhood in an undeveloped consensual world for discriminant exploitive reasons, and this is why you learn in school whatever it had been invented as a theory hundreds of years ago, as nonexistent forces of the universe along with elementary particles meant only to launder money for the Families in control of entire nations. While as stated, it is easier to notice the entire harmful consensus taking place in Russia, old Germany, and North Korea, while it might be surprising to see it in the West and in the East. While the consensual Brotherhood does a good job while serving the Families of the upper society, by maintaining order, decay, and austerity within the Masses and the Brotherhood, at all costs.

How much does it cost to save the world from shortages and from the entire austerity? This question never has an answer, because austerity itself is deliberately maintained in the world, as it enhances ignorance, misery, shortages, loss, lack of fulfillment, discrimination, exploitation, and eradication, while without these, nobody serves anymore, making all tyrants impossible. Wars, medicine, and corrupted bureaucracy are part of the same continuous austerity, harming the world similarly. While the Brotherhood implements everything and it works very well, because misery and death are very efficient, while this is the same Brotherhood that was meant to look after the world and actually save the world. While with no one to save the world, everybody dies, Masses and Brotherhood alike, with the Elite hiding underground from one last year or so. Good luck to the world, since it happened before, in this exact scenario, and everybody died shortly, Masses, Brotherhood, and Elite, while you can still find their impressive caves throughout the world, where they had been hiding and dying shortly.

There might still be accurate theories invented this entire time, decades and centuries in a row. Mainstream science cannot be more deceiving, while 'accurate theory' is not exactly a correct statement, since all scientific theories are speculations but not exactly accurate, since the word 'theory' means speculation. Therefore, all scientific theories are only speculations, and you are capable of countless of speculations yourself throughout the day, while no one is actually accurate. It is the same with scientific laws, since laws only define consensual knowledge, accurate through law, law that is accurate only within the specific jurisdiction where it is implemented.

Yet there is more to consider, because higher laws implemented consensually throughout higher worlds become accurate natural laws by default in the created worlds, because the higher worlds form, hold, and maintain all their lower worlds. This is the case with all natural laws of the universe set in place at the base of all created worlds by their creators themselves, in any manner they choose alone or alongside others in a higher consensual manner.

However, when the people of the lower worlds decide consensually upon their own laws in their own lower worlds, they do so only in a lower, erroneous, consensual manner, because they cannot chance consensually the natural laws of their own lower worlds, since they are already set in place by their own creators whenever these created their worlds. However, people can always turn their backs to their creator by choosing to consider erroneously everything they desire, even that this entire world started with a big bang, that it has additional elementary particles and not only those made possible by their creator, and that all people should be arranged minutely within very specific social hierarchies in order to assure a continuous exploitation and eradication, because only in this manner, tyrants become deities but only consensually, and are served and venerated accordingly by everybody else holding this consensus.

Therefore, you cannot form a scientific consensus with Einstein and Hubble to decide upon the creation of this entire real world by an entire big bang, because this world had already been created normally and statically, not through a big bang. How many times do you create your computer worlds or your reveries with a big bang? Never, but you create everything normally and statically.

This is the difference between consensual knowledge and accurate knowledge, because you can always have a group of people deciding what is meaningful and necessary in the world, yet it is not reality, but only consensual knowledge. As it is not exactly what the world truly needs, but only what it needs consensually, whatever authorities decide that it needs, as it always happens. If authorities decide that science should focus on elementary particles, big bang, evolution, and universes of thirteen dimensions curled up on themselves, now this is what science researches and this is how it spends its funds, while the world suffers in shortages, ignorance, darkness, servitude, and indoctrination. How exactly does the neutrino make a better world? Ask Pauli, and he might invent an entire universe filled up with neutrinos and meant to balance the equations of this universe, including its lack of food.

This defines the human reality forcefully, whatever these specific groups of people decide consensually to be the human reality, and not the actual human reality. If you want to know more about the consensual human reality, you can study all mainstream and alternative knowledge, because everything is in education, in science, and in the media. While if you want to learn more about the actual real human reality, it is not science offering it, since as it states itself, science is only consensual in nature, based on its own scientific consensus, but not accurate. While the alternative science does not offer an accurate truth either, since alternative science offers you also consensual knowledge about the human reality, knowledge that is agreed upon by an alternative group of people, while this is not accurate either.

Similarly, all jurisdictions and all ideologies are consensual in nature, offering all their beliefs and laws that remain compatible only with the Consensual Matrix, and not with the real world. Therefore, these can offer only the consensual human reality, along with the mainstream science and alternative science, but not the actual real human reality.

It even seems awkward to divide the human reality into the fake consensual human reality and the real human reality, yet since this is the case currently in this world, this is what we must study, otherwise you learn only the fake consensual human reality, exactly as the current mainstream and alternative science teach it.

While as you study this circumstance closely, you notice how currently, the fake human reality spans the world, while the real human reality is censored and removed systematically from the world, by the people themselves, by the entire humanity, in a continuous consensual conspiratorial persistent manner. While if you ever seek to study or even teach the actual human reality, they harm you badly, because they never want anything real, intelligent, accurate, and alive, but only everything consensual and conspiratorial, since only through consensus they can have drugs and tyrants in this world.

Where can we find all the knowledge about the real human reality, needed to complete our model of the human reality? Not in everything consensual in the world, as all ideologies, mainstream science, alternative science, media, social domains, and the rest of the jurisdictions of the world, since these are consensual in nature, and never offer actual accurate knowledge. We find it nowhere, since the human society is currently consensual, and cannot offer anything compatible with the real intelligent living human nature, but only consensual knowledge. With everything else about the real world made erroneous deliberately, as the big bang and the entire particle physics capable to derail your understanding of the real world by default.

Elementary particles should stand at the base of this world, as science claims. Yet if these are not true, we have to find our own way while modeling this world, and it might become tedious. Yet we model the entire human reality in this book, not only the real and consensual world. While by not being able to understand this real world along with the entire wider world, we might not be able to understand everything meaningful about the human reality either, including the human society, human mind, human knowledge, human behavior, human nature, human dreams, human fulfillment, human meaning, human lifestyle, human rights, human development, and human interconnectivity.

They use one force in places as Area 51, since it is always accurate there, but they use four forces for the rest of the people of the world, to make things more complicated and to keep a million scientists busy in the world throughout meaningless research, away from anything important as free of cost energy, better transportation, and better living conditions. While these scientists love it in this manner, since they get to show to the world how wonderful they are in the world while dealing with elementary particles, dark energies, and forces that only they can understand and tell apart but that have no standing and no use in the world, no use for the people, for the living human beings. Since this is the difference between the consensual and the real, one addresses only corporations and jurisdictions, while the other addresses only living beings, with living human beings included.

Currently in the mainstream or in the alternative, you cannot study and understand the human reality because the consensual human reality is available excessively everywhere and it is mandatory to consider, because the current society is consensual entirely, providing everything consensual to all human beings, as all laws, ideologies, jurisdictions, jobs, benefits, wars, military, courts, jails, documents, lawyers, licenses, permits, ID cards, social databases, systems of justice, bureaucracy, money, banks, loans, investments, business, profits, education, addictions, servitude, jobs, lodges, theories, authorities, insurance, political parties, votes, presidents, entertainment, orders, registration numbers, commerce, news, privileges, indoctrination, Internet, corporations, social media, and trademarks, everything consensual is provided abundantly in society, since the current society is consensual itself and meant only for corporations, not for living human beings, while always considering you a corporation, not a living human being.

This is why there are people dying of misery, sickness, and shortages in the world, because it is not exactly an incompetence of the current society to help them and provide to them in any manner, but the current society does not even apply to them since they are living human beings, while the current society is consensual, meant only for corporations, while corporations do not eat. This is why humans have to find on their own all their resources to assure their subsistence, intelligence, development, meaning, interconnectivity, and fulfillment, in private and during their own time, apart from the current society, because the current society does not even apply to the human life, since it remains incompatible to living human beings altogether, since it is consensually dead. They also killed all real nations of the world, replacing them with jurisdiction corporations that only have the names of the old nations but written in uppercase letters, for people to have nothing providing to their actual life, intelligence, development, meaning, and fulfillment, but only consensual laws, applying only to their consensual corporations. While it was supposed to be the actual goal and priority in any intelligent human society to provide to the natural needs of humanity, because humans are intelligent living beings, and society was supposed to provide everything natural necessary for the human life, development, consciousness, and natural interconnectivity. This is why the current society never provides anything material, natural, and necessary in the world, but only everything consensually necessary in the world, as the entire bureaucracy, and you can tell the difference. It is so perfectly agreed in this manner through the Elite and the Brotherhood, that you cannot even find the necessary accurate knowledge about the actual, real, natural human society and entire reality, but only about the consensual human society and reality. While now we can certainly tell the difference between the two human realities, real and consensual.

This explains all deliberate errors in science, because the current science applies only to consensual theoretical universes that can always be in any manner authorities agree and decide that they are, but not to the real natural living world, the entire real world. This is why, the more you study the consensual big bang model of the universe, the more erroneous it looks from an actual real perspective. Yet all these lies forming it help us understand why they have been set in placed, by who, who ordered them, what stereotypes they maintain, who profits, and what they hide, while leading us to see and understand the real shape and meaning of the real universe. Furthermore, understanding this scientific consensus helps us understand better not only the current consensual science, but society, religion, and ancient history, among others, everything necessary to model the actual, real human reality, and not only the consensual human reality that you already know well from school and from the media.

Science works hard to keep this consensus valid, and it does so flawlessly, with hundreds of millions of scientists, teachers, and office workers in perfect conspiratorial agreement that involves everybody in the world. Because without a consensual model of the universe at the base of this entire world, humans might see the truth through lies, that it is actually their own world, while they are actually living human beings, and might start fulfilling normal human needs and meanings instead. While in this manner, no one can control them anymore, making all tyrants impossible.

What consensual human reality exactly? No one tells,

because no one actually understands too much about the current Brotherhood and the entire Consensual Matrix, since all knowledge is modulated and enclosed at all levels. While these elementary particles do not even exist, they are only field, only electric, magnetic, and electrostatic field found in various states. This is important to research, since there is nothing rigid and material at the level of all elementary particles to be even called particles. Here is where relativity unifies with classical physics and with electrodynamics, all three being accurate physics domains, and not exactly particle physics and quantum physics, which are impossible and irrelevant in the real world.

For the neutron decay, there have been barely detected a few neutrinos throughout the century, in dubious circumstances, as during a holiday when no one was present. While in theory, neutrons decay by the zillions, everywhere, so there should be zillions of neutrinos and antineutrinos everywhere right now, ready to be detected. Yet they are not, since they are bogus, or consensual, fiat, not there, along with the neutrons and antineutrons generating them.

Where is the error regarding the neutron decay along with every particle decay? The error is that physicists were supposed to use the law of relativity for particle decay, and not classical mechanics, because the energy density in the field of all elementary particles approaches relativistic states, therefore becoming relativistic circumstances, forcing you to use the law of relativity for all particle decay. The law of relativity is a natural law of the wider world, a supreme characteristic of Life, and a natural law of this world, because our Creator placed it at the base of this world, along with the other natural laws of this world, when he created our world.

This is the real law of relativity, not the two theories of relativity stolen by Einstein, which are something else. Einstein's two theories of relativity are very simplistic, very enclosed, and mostly erroneous, yet still posing as the natural law of relativity from the base of this world, even though the two theories of relativity are called theories, yet you are forced to use them as natural laws, in a dualistic manner, both real and consensual. You also notice how the current consensual science forces everybody to disregard the natural law of relativity that would help everybody to understand Life, existence, reality, this world, and the entire wider world, while forcing everybody to consider only the two restraint and erroneous theories of relativity made or stolen by Einstein, for the reason that Einstein is the most capable physicist ever.

Dualities are found everywhere in an undeveloped consensual world, even between the natural law of relativity and the two current consensual theories of relativity. Similarly, Kim is a wonderful man, the savior of North Korea, while he is a cruel tyrant megalomaniac, the destroyer of North Korea, in a continuously impossible duality, both real and consensual.

Then why don't they use the two theories of relativity to model decaying elementary particles, for better results? Because if you do so, you state implicitly that neutrinos and all leptons do not exist, along with the weak and strong forces of the universe, along with the big bang, and along with the entire dark energy and dark matter. The decaying elementary particles and the universe as a whole are not considered relativistic circumstances, because the current theories of relativity apply only to the orbit of Mercury, to black holes, and to spaceships traveling close to the speed of light, while disregarding everything else. Therefore, nothing else can ever be in a relativistic state according to the current theories of relativity decided by Einstein, while the theories of relativity considered before Einstein were merging gradually towards the actual law of relativity governing this world.

Because the theory of relativity, even though it had been first formulated by Poincare, Larmor, and Lorentz one decade before Einstein since it was common knowledge at the time of Einstein, ended up truncated, contorted, and derailed, since it was advancing gradually towards the actual natural law of relativity. Relativity itself was truncated, contorted, and derailed. Instead, now you have Einstein with his theory of relativity, which is not the same. Because as you already know, theory means only supposition or speculation but not the real thing, not relativity, not the actual real law of relativity. Who exactly is the greatest physicist in the world? Einstein. Why? No one knows exactly why, but yes, it must be Einstein, no? Since this is the scientific consensus, with the name Einstein on top of all lists of references. More precisely, they truncated, contorted, censored, and covered up an entire natural law of this world, only to keep the human consensus instated, and only to make possible all tyrants of this world, throughout yet another dark age of this world. While you can never notice it if you do not study the human reality accurately in all details, while differentiating it from the consensual human reality.

There was a law of relativity long before Einstein, similar to all the other natural laws of this world, since the law of relativity is a natural law in itself. Yet the invisible kingdom derailed it into the two theories of relativity made or stolen by Einstein. While these two theories of relativity made or stolen by Einstein apply on only three very particular circumstances: spaceships travelling at speeds close to the speed of light which will never happen anyway, star light bending around Venus which is irrelevant compared to the price of rice in Bombay, and very specific circumstances related to black holes which are always unreal because black holes do not exist. You can have protons, pulsars, and neutron stars but not black holes, because according to the actual law of relativity, you cannot have mass density exceeding the mass density of protons, pulsars, and neutron stars.

The law of relativity is capable enough to explain everything in the universe alongside the rest of the laws of the universe. This is why these laws of the universe are discovered, studied, modeled, and explained throughout the entire classical physics and more, since they are the most accurate scientific knowledge ever, because they are placed at the base of this world while defining and determining all events and all lines of causality of this world, otherwise this entire world cannot exist. This is the case with all natural laws placed at the base of all worlds and realities, which can be different or similar from one reality to another, yet by being at the base of all worlds and realities, they define minutely all events and all lines of causality of all these worlds and realities.

However, currently, you cannot use anymore the law of relativity in physics, even though the law or relativity itself is at the base of our world, because Einstein himself forbids it, with his two theories of relativity replacing the actual law of relativity consensually, while contorting and truncating the law of relativity itself to the only three irrelevant circumstances stated above. What exactly can you do with these when you have to model and explain an entire universe as we try to do in this book? Nothing at all, but you end up with neutrinos, big bangs, dark energy, black holes, and black energy instead. Yet since these neutrinos cost trillions of dollars to detect, they must mean something.

How exactly does Einstein feed the world through his two bogus theories of relativity if they apply only to Venus, Mercury, and black holes? He feeds himself and the invisible kingdom, while derailing physics entirely, since now without the true law of relativity, good luck to everyone while studying anything relativistic. Is food ever relativistic? Then what exactly does it take for scientists to study food itself, in order to feed the world? Do they ever study the most affordable sausages in the world, in order to feed the world? No.

As stated, there is a difference between Einstein's theory of relativity and the actual law of relativity discovered and developed decades before Einstein. The theory of relativity is only a theory or speculation, used only in three specific circumstances, while relativity is the actual physics, which must always be used under all relativistic circumstances, including cosmology, electromagnetism, particle physics, electrodynamics, optics, and astrophysics. Yet Einstein was in the invisible kingdom, serving minutely the invisible kingdom, at the time when they took over the world from the old European aristocracy, while replacing the religious ideology used by the old European aristocracy to rule the world with Einstein's and Hubble's cosmology and with the entire consensual science, the new ideology meant to rule the world in place of religion.

Creationism versus the big bang, both used to rule the world, one dark age after another. Some people state that the current science is the new religion of the world, which is still erroneous, because while religion is a religious ideology, the current science is a scientific ideology, based on a scientific consensus but not on accurate science as it should.

The current science is not a new religion even though it seem so, yet both religion and the current consensual science are ideologies, always incompatible with the intelligent human nature, and always harming humanity by keeping it in dark ages, since they are both dogma, and dogma ruins the world. Both religion and the current consensual science are ideologies, and they are capable to rule the world in a contorted consensual manner, because the current science has priority in court, while in this manner it holds the entire authority in the world. It was similar in the previous dark age, when religion had priority above all authorities, and it ruled the world accordingly on behalf of the old European aristocracy, similar to the current consensual science that rules the world on behalf of the invisible kingdom. Yet the invisible kingdom is worthless, and it already lost the world to the dictators of the East, who also use consensual science and the entire consensual system of justice to rule the world.

Relativity, as it had been first formulated before Einstein and before the entire invisible kingdom, was meant to explain everything about this world, along with classical mechanics, mathematics, and electrodynamics, alongside all natural laws of the universe. With many of them capable enough to open a window of understanding beyond this world, meant to define all realities, and to allow the understanding of this world by contrast, of the field engulfing it, and of other realities, being key part of the entire model of the actual human reality. Yet not anymore, because the current theory of relativity the way Einstein truncated it deliberately replaces the law of relativity entirely. While the theory of relativity is divided in modules, and must be used under very narrow circumstances, only as it is stated by Einstein, for black holes, space ships travelling at relativistic velocities, and for the curvature of light rays in the vicinity of very large space objects. How do these feed the world? While in contrast, the actual law of relativity applies to all relativistic circumstances, similar to classical physics, mathematics, and electricity and magnetism.

If you were wondering what Einstein did for the Consensual Matrix so important that he is venerated endlessly, this is what he did, he derailed the human knowledge so badly through his theory of relativity, that no one can figure out the actual human reality anymore, being forced to use knowledge about the consensual human reality instead, and in this manner, being forced to exist in the Consensual Matrix instead, but not in the natural, living, real human reality. This makes a tremendous difference in the world, while making all tyrants possible, along with the entire consensual Brotherhood serving them closely, while Einstein certainly helped, along with Hubble, Pauli, Heisenberg, De Broglie, Oppenheimer, and the rest.

Do you see how the scientific consensus not only controls the output of knowledge by controlling the scientists themselves, but it manages to censor common knowledge through specific scientists as Einstein who managed to contort an entire domain of physics and make it apply only to three insignificant circumstances: spaceships travelling close to the speed of light which is science fiction, black holes that are always nonexistent, and rays of light bent by stars and planets, which is actually trivial, irrelevant, useless knowledge coming with price tags of trillions of dollars. People believe everything through lack of options, while ignoring the entire real world, yet if you ever ask them who the greatest scientist in the world is, they answer immediately Einstein, while cheering for keeping them ignorant.

Einstein took over relativity decades after it was discovered, with one reason, to patent it in his name, and therefore to be the one deciding when it should and should not be used. This is the case with many inventions, theorems, and discoveries, since you cannot use them legally under any circumstance if they are already patented, or you break the law, while all patents are consensual. Yet real living people starve for this consensual reason, because the specific technology meant to feed them is locked away in some drawer and you are not allowed to discover it, to invent it, and to use it, because it is already patented and you break the law if you do. Do you see how the entire Consensual Matrix manages to take over real life while constraining living human beings to obey it, just by depriving them of everything necessary to subsist and to be alive, including the necessary knowledge capable to feed, shelter, and transport them?

Because Life and our Creator gave you this knowledge in the first place, but you refuse to use it in order not to break the consensual law. While this is your choice in life and in the world, because you are free to do as you please. Yet do not expect more knowledge coming from Life and from our Creator to use and to feed yourself, because accurate knowledge is unique, and there is nothing else. There is only one relativity in the world and nothing else, since knowledge is unique, while you should be more responsible with everything that you decide, consider, and do, because others take everything away from you, and you cannot use it anymore.

Relativity itself might not feed the world directly, yet you can always use the natural law of relativity to generate relevant knowledge as free of cost energy, while this specific knowledge can feed the world freely or at very low costs, eradicating all basic shortages. Even eradicating money in the world, since with free of cost fulfillment of all human needs, who exactly needs money? Yet tyrants are impossible without servitude, currencies, and price tags, since this is why they remove systematically all intelligent human knowledge, while replacing everything with string theories, universes of sixteen dimensions, and black holes. The consensual removes and replaces the real, the accurate, the intelligent, and the alive.

Because when you control the food, you control the slaves. Yet without money itself in the world, you cannot enforce servitude anymore in the world, and all tyrants of the world have to descent from the top of the world to tend to themselves. Which is not exactly what the tyrants want, which means that it is not exactly what you have.

This is why people cannot use the law of relativity, because it is censored away from the human knowledge for various consensual legal reasons. Yet if you happen to discover the law of relativity yourself, or if you happen to invent any device based on the law of relativity, capable to provide free of cost energy and therefore free of cost necessities in the world, you cannot publish it because it is already patented. If you are in the Masses, nobody listens to you, while you cannot even publish your research. While if you are in the Brotherhood, you should have never discovered it, since as a scientist Brother, you are meant to research only what you are ordered to research, big bangs and universes of sixteen dimensions curling up upon themselves.

Furthermore, if you are not a scientist, it means that you are in the Masses, and therefore you are free to research and invent or discover anything you please, yet everything that you invent and discover as a nonscientist is considered irrelevant in itself, it remains outside the scientific consensus since you are never in the approved lists of reference, and it is never accepted in the world, by default, because the scientific consensus has already locked away all similar important human knowledge. This is how science and its scientific consensus maintain the monopoly over the human knowledge, and it is going to be in this manner indefinitely.

Everything relates with the main social actors in control of science. You can identify them yourself, since they are of the same genetic background, they are the invisible kingdom, and this has nothing in common with religion. These people controlling science control the entire Western Civilization, and through it, they control most of the world. These people are genetically related because they had their own nation in the Caucasian region in Asia, Georgia, but they migrated in mass to Europe, over half a millennium ago and longer before, and so they took over everything, including finance, education, Internet, entertainment, science, media, and business. The world is theirs for over a century or two, since they managed to take control of the old Brotherhood, while using it to control everything in the world, including science. Yet the current dictators of the East change the world order once again, while taking down the invisible kingdom.

The real world should never be consensual, but real, human, intelligent, and alive. In contrast, laws, rules, patents, and ownerships are simple agreements, part of the consensual human reality, agreements between you and them. If you refuse to have anything in common with these agreements, you do not have to respect them, and you can invent and discover whatever you please. Because knowledge is free in the world, it is part of this world, it is part of the natural laws of the universe and they are free by default, since these same natural laws define you as a natural living human being. This is the case with the classical physics, electrodynamics, relativity, and mathematics, since these are also part of the natural laws of the universe, while no one patents the mathematical laws of addition and multiplication, to have to pay royalties every time you use them or any time you use the conservation of momentum in high school and throughout life. You can patent books, paintings, or technology if you wish, but the knowledge behind them is always free to use, as it is invented and lost repeatedly throughout the zillions of civilizations of the wider world, with no one having to pay royalties to the science of Earth.

This is the case even if you use the natural laws of the world in order to help the world, since there is sufficient technology to allow electric cars to replace combustion engine cars entirely, but it cannot be used because it is already patented and locked away, and therefore you break the law if you do. This is why Einstein stole relativity, in order to truncate it and to call it a theory, the theory of relativity, to patent it, and to put a lock on it. This is why all relevant theories, principles, and laws in science are stolen and patented away, so they cannot be used in the mainstream of science, in the exoteric science, in the science for the Masses and the Brotherhood.

Consequently, tens of millions of scientists from the mainstream science have to use bogus and irrelevant theorems, laws, and principles instead, and this is why they obtain results as thirteen dimensional universes curling around every point of the universe, or impossible explosions as the big bang itself costing trillions of dollars. This is why they laugh in the media when they come up with these theories, since it is a prestige for many Brothers to keep the humans of Earth ignorant, confused, diverted, controlled, and exploited. While the genetic lines of Earth die in large numbers within the Masses and the Brotherhood, leaving behind only the tyrants and dictators of the world.

4 CONSENSUAL THEORIES AND THE ENTIRE SCIENTIFIC CONSENSUS

Can our world still survive with flawed and altered laws and theories of science? Can an entire civilization survive with these major scientific errors and restrictions in place? Yes, yet it forces science to a dead end, with all scientists Brothers proud to take it there. There have always been errors in science, from the theory of evolution to the theory of relativity and the big bang theory, and nobody died. What people did, they formed a consensus to consider unanimously these assumptions true, consensually true. This is how they lived their lives while changing the entire world order in the process, on behalf of those above who managed in this manner to have an entire world for themselves. While in science, this is not intelligent reasoning anymore, but ideological thinking, based on beliefs and stereotypes, which is similar to what you have in religion. Therefore, ironically, science and religion finally have something in common, consensus, beliefs, control, dogma, hierarchies, and ideologies, since both science and religion are ideologies, after all the fight that they had and still have over evolution and creationism.

Can't we just get along? Yes, certainly, since the consensus allows it. This is the difference between the accurate and the consensual, because while existence can define only the accurate truth as accurate, the consensual existence can validate anything to be true, but only consensually, even big bang and creationism, if there are people willing to validate these consensually. This is how all mainstream ideologies get along, as part of the consensual human reality.

Yet is this accurate? No, yet it makes a profit, and therefore it agrees with the major ideologies governing the current human society, consensually. While it is typical for ideologies to fight in this manner, as they always do, for supremacy, uniqueness, and recognition.

While at your lower social level, when there is a consensus you stick to it, since it makes your life easier, regardless of what you are made to do to the world since everybody harms the world, and no one blames you. This is how you end up fulfilling artificial needs and meanings in the world on behalf of the Consensual Matrix, and not natural needs and meanings for Life, while harming the world. Otherwise, you lose your job and you never find another one in the scientific field, but only in a restaurant or in a grocery store. This is how you obey everything, and if you have to invent an entire family of bogus elementary particles to balance your equations and stick to the official model of the universe, then this is what you do, you invent everything exactly as told, while everybody applauds both your ingenious idea and your great obedience to the Brotherhood and to the scientific consensus. You save them through your bogus science, and it is called intelligent prostitution.

You are not the only one, because it happened before as it happens repeatedly throughout the consensual science, with the electricity magnetism duality, the dark matter invention, the particle wave duality, the inflation of the universe invention, the strong and weak invented forces, the big bang speculation, the with the invention of all quarks and all multidimensional universes.

Scientists invest their efforts towards finding new methods

of detecting neutrinos, since if they ever turn their attention to question the accuracy of the laws used in the neutron decay, the law of conservation of momentum, they never get funds, or this is what they say. Others had detected similar errors and found similar answers, yet you never hear about them. Their work can never be considered accurate, and it can never be carried forward and stated in references, since it is not part of the scientific consensus, never accepted in the mainstream of science.

The consequences are major, since the current Physics had contorted to a point where you cannot take forward any scientific idea, since nothing relevant makes sense anymore. Everything already looks impossible, out of this world. If you insist to discover new theories, you have to improvise consensually the way Pauli did a century ago, with his invention of neutrinos and antineutrinos.

Scientists had improvised as best as they could with the big bang theory and the age of the universe. Hubble had to adjust H, his famous cosmological constant so many times, because his age of the universe always came up to be too short, making the entire universe younger than continents on Earth or than the Sun itself. This is a major discrepancy between consensual truth and accurate truth, yet everybody accepts all scientific knowledge as accurate knowledge, and the world goes on, consensually.

Hubble changed the age of our universe many times throughout his life to give enough time for Earth to form, and then for the Solar System to form. While all physicists followed his speculations blindly, in order to keep the consensus going. The age of the universe is settled currently at thirteen billion years. Thirteen point eight, yet this is impossible, since even stars and planets are older than this, which means that all blue and green stars are older than the entire universe. There are rocks on the Moon older than that, if the Moon, the stars, and the entire universe are actually there and not only lights in the sky as seen from Earth. Our own galaxy must measure its age in zillions of years, and it is a young one. Imagine the age of our super cluster of galaxies, or of the entire universe, if these are objectively real in the sky as they seem, and not only lights.

When you use classical mechanics for systems as large as the universe, you obtain erroneous answers, since this is another relativistic circumstance, involving not very large mass densities or very large velocities, but very large scales, distances comparable to the size of the detectable universe.

Why is this case not covered in Einstein's theories of special and general relativity? Because even if these two theories of relativity had won Einstein a Nobel Prize each, they are both empirical and not intelligent, meant to explain only three circumstances: the distortion of the orbit of Mercury, the bending of starlight in the vicinity of large space objects, and the distortion of time at relativistic speeds. When you make an important model in physics, as the theory of relativity, you must master it entirely and understand every detail, yet Einstein did not. He was not even able to explain his special theory of relativity when he first formulated it, over a century ago, since the idea was not his in the first place. The formulas that he used, the formulas that make him famous, the greatest physicist ever, are not his. Larmor had transformed them from Maxwell's Equations from electrodynamics, then Poincare fine-tuned them from Larmor, then Lorentz took them from Larmor and Poincare, to name them Lorentz transformations, not Einstein equations, since Lorentz had them first or simultaneously with Poincare and Larmor, while Einstein was not even a physicist at that time. Because relativity used to be common knowledge long before Einstein, just as long divisions, friction forces, and the conservation of energy. Einstein stole all relativity equations and used them as they were, along with all major ideas associated to them.

More precisely, the invisible kingdom forged everything, as it altered and contorted the entire human knowledge, while stealing the entire world in the process. However, the invisible kingdom is worthless and did not have artists and physicists, and therefore they made Einstein a physicist, and placed in his name the entire relativity, but only as a theory, while using the formulas and equations that were currently in use, they published everything in the name of Einstein, yet they published everything one decade back in time, while forging an obscure advertisement newspaper from one decade in the past. Furthermore, they gave Einstein two Nobel prices, in order to place him at the top of all scientists of that time, as an authority of knowledge.

Why are the most important scientists currently part of the invisible kingdom? Who owns the human knowledge currently? The invisible kingdom. This is important, since once they own science, now science claims that all native European genetic lines are actually Caucasian, coming from Asia, giving the entire world to the invisible kingdom. Therefore, you have to mark yourself as Caucasian on all official documents even if you native European, and in this manner, you become a servant of the invisible kingdom by law, knowingly or unknowingly. While you ask for it continuously, legally and with registration numbers.

Currently, officially, the relativity transformation equations along with all relativity ideas are associated exclusively to Einstein. Einstein himself had introduced the invariance of the speed of light in every system of reference and the space and time distortion by relativistic masses and velocities, yet Einstein was incapable to explain and prove them, since he only stole them. Not only this, but the use of relativity in any other domain is banned, since it is called the theory of relativity, which is the speculation of relativity, and it is significantly distorted and truncated. This is how humanity's means of perception and understanding of this world and of the entire wider world remain unavailable, after this entire time, and will always be unavailable, keeping humanity in its dark ages.

Gradually, other physicists gave the necessary proof for Einstein's theory of relativity, as best as they could, in order to be noticed. They managed to prove that time slows down at relativistic speeds by using simple geometry. Einstein proved why light curves slightly while passing near the Sun using the theory of relativity, and how and why the orbit of Mercury is affected by the large mass of the Sun, yet it seems that others did the same before him, and they are now forgotten.

Science decides everything currently, consensually, yet not exactly accurately. As a reference, the great inventor Edison discovered the light bulb and enlightened humanity ever since, while everybody knows this. However, do some research, to find seventy light bulb patents registered long before Edison. Seventy. Because it seems that the accurate truth never matters in the Consensual Matrix, but only the consensual agreed truth, and it is not the same thing. While without an accurate truth, you can never have an accurate model of the human reality, but only a consensual one. This is the consensual human reality, and it is not the same. If you want a model of the consensual human reality, you can study mainstream science with its own big bang theory, theory of relativity, theory of evolution, and dark matter.

Why contorting the scientific knowledge itself, the knowledge necessary for an accurate, imperative human development, at the level of the entire human civilization? Because those were imperative times. It was the time when the old world order changed, when it was replaced by the new world order, which was the new world order of those old times, during and after the two world wars. Not only the Elite changed then, almost entirely, from the old European royalty to the Rothschilds of the invisible kingdom, but all previous social orders, social exploitation, and social agendas did, replaced with the new ones, which are the current ones, along with all social actors of those times and all their hierarchies, genetic lines, laws, and bank accounts.

To make room for the invisible kingdom to take over the world entirely, and so it did. Is it a coincidence that the dictators of those times and decades after managed to eradicate hundreds of millions of people, all part of very old genetic lines of Earth? What was the population of Earth? Three times larger than that? While this is called genocide at a massive scale, and it still goes on currently, through wars, terminal illnesses, starvation, pandemics, sterility, enforced abstinence, genetically modified viruses, abnormal sexual behavior, and abortions.

We notice that there is more to the human reality than a simple scientific consensus. There is even more than a pattern of similar consensus everywhere, in all social domains. There is an entire pattern of occurrences taking place repeatedly throughout history, affecting the entire world. We have consensuses, patterns, precisely altered scientific knowledge, and preordered laws and ideas enforced for the lower social classes to learn and follow. There are white lists of predefined scientist to consider and follow with everybody else to ignore, and there are lists of predefined results to obtain in every social domain, in a highly precise agenda that everybody follows unconditionally. This is the consensual human timeline, consensually altered as you can distinguish it in the world, forming the duality between the consensual human reality and the real human reality.

While as a scientist, you have to follow the consensual human reality, harming the world every time you do, yet you never care, probably because you have already burned your care at the lodge throughout rituals. While with the old genetic lines of Earth targeted directly and almost extinct, everything is part of this massive revision of the entire human reality. Just study your own consensual needs and meanings as you fulfill them diligently for the Consensual Matrix, to see how you are in all these just as well, as the rest of the world, harming yourself and the rest of the world in the process, while marching merrily to the common grave along with the rest of the world.

Science is not free, science is not accurately true, but science is an ideology, the way it was a decade ago, last century, and a millennia ago, matching each world order closely. It is nothing else, nothing more, nothing special, but a normal scientific ideology. There is no thought associated to science, no reasoning, but a simple belief in scientific theories, principles, and ideas to be true, the way it has always been throughout history. Our times are not privileged, the best times to be alive, the most advanced, and neither are today's people and their leaders.

This means that science does not work hard to find the accurate truth, the way it was supposed to do, but science works hard only to answer and satisfy people's demands for knowledge and truth at the current level of their knowledge and development and on their current circumstances, regardless if these scientific answers are true or false. This gives science the legal freedom to offer the Masses not an accurate model of the universe, but a model more appropriate and more adequate for people's status and understanding of the universe, as the big bang theory itself. The big bang is not accurate since it is only a theory, not an actual real event or a natural law, and therefore it is not part of the actual human reality, but it is part only of the consensual human reality.

Now we can understand better what happens in the world of science, and what happened with Poincare, Pauli, Einstein, and relativity. The theory of relativity is not genuine, but reverse-engineered. It had been created only to explain to the Masses a few anomalies observed by the astronomers of that time, as the distortion in Mercury's orbit and the curvature of light near the Sun, while accurate scientific models should apply to all similar circumstances, and not only to three cases. Einstein himself could not understand his own theory, because his model is not intelligent, but empirical. It is the other way around, his relativity model is empirical, not intelligent, because he could not understand it to offer it as an intelligent model in the first place. The implications are that the theory of relativity today is not complete, and this is why it can be scientifically applied only to a limited number of cases. The transformation equations from the law of relativity were supposed to be used in the conservation of momentum and energy for the decay of the neutron and of all elementary particles, they were supposed to be used in cosmology to describe a real model of a stable universe, and it is supposed to work flawlessly in electrodynamics and everywhere else.

Decades earlier, Larmor, the first discoverer of relativity, started to use his own principle of relativity in a model of an orbiting electron within the atom. This model was not good enough according to the current consensual science, because 'he did not get the velocities quite right.' Note that Larmor implicitly considered the atom to be a relativistic case, since this is why he used his theory of relativity to model it, and if continued, his model could have led to a true model of the atom, a true model of the nucleus, and implicitly to a true model for the neutron decay, eliminating the need for the neutrino adjustment. He might have done everything correctly and obtained accurate results, since it is not too hard to form the right equations, yet the current science censors his work. Coincidentally, Larmor had used the term 'electron' many years before Thomson, since he had clearly identified it, while his mathematical model of an orbiting electron came very close to the scientific consensus, but not close enough according to the same scientific consensus, so they praised Thomson instead.

Why this entire effort to censor an old scientist? Just study the dates. Larmor was not part of the invisible kingdom, as he used relativity and did everything during the old world order, during the old European aristocracy, and therefore long before the invisible kingdom, while the invisible kingdom had no artists and scientists at that time, since the invisible kingdom is worthless. Einstein was actually part of the invisible kingdom, and he 'invented' the theory of relativity much later, during the new world order, during the invisible kingdom. More precisely, Larmor was supposed to be the Einstein of today if it was not for the invasion led by the invisible kingdom during the two world wars to change the world order on their behalf. Therefore, now it is Einstein instead, not Larmor, since he is part of the invisible kingdom, and the invisible kingdom needed artists and scientists, since it is worthless.

Larmor was ready to apply relativity to everything relevant in the world, forming the actual model of the human reality that we seek in this book, while taking the science of those old times further than the current consensual science. Could it be that the science of those old times was already more advanced, but everything is erased currently, revised from the human knowledge, and you will never know it? Real science and accurate intelligent human knowledge are not compatible with the Consensual Matrix since they make tyranny impossible, and therefore this is why it was not allowed. This is why the Consensual Matrix changed its freight horses on Earth then, from the old European aristocracy to the invisible kingdom, since it could trust the invisible kingdom more. While currently, the Consensual Matrix trusts the dictators of the East even better, placing them in complete power instead.

Who exactly is Larmor? Larmor is nobody according to current science and according to the invisible kingdom, probably because Larmor is not part of the invisible kingdom in the first place, and this is called discrimination. While the invisible kingdom stole all his work to profit from it, and this is called theft. Yet for the invisible kingdom it is called profit, and it is actually a cultural pride, just study all ideologies of the invisible kingdom to see it for yourself. While the act of theft is irrelevant from the perspective of the invisible kingdom, because the rest of the world is not even made of humans according to the invisible kingdom, but of animals, cattle, and abominations, since this is what they name the people of the world if they are not in the invisible kingdom, which is discrimination.

This defines the human reality currently according to the invisible kingdom, and it is significant to know it. Science already states that you are an animal and not a plant, and therefore you even accept to be an animal.

Is Einstein to be blamed for his poor understanding of physics? Einstein, the greatest physicist ever? No, they should all be blamed, all scientists, along with their entire consensus. Yet what if Einstein understood everything, and already made an accurate intelligent model of relativity and of the entire objective world? What if all physicists were allowed to use his relativity model in all relativistic circumstances, leading to a true model of the world? Then science would have already provided all the necessary knowledge to make this world a proper intelligent human place, with the Consensual Matrix on Earth already departed.

Yet this is never the case, as you can notice now, because it does not matter how close humanity comes to achieve its freedom and natural independent living, because something always ruins all intelligent human achievement, to erase the entire intelligent human environment, and to return humanity to its dark ages, back in the Consensual Matrix. Because the Consensual Matrix is highly capable to control the timelines of entire worlds and realities, including the human timeline.

It is important to identify how it does so, because just by acting consensually, the humankind will never be able to free itself from the Consensual Matrix in order to achieve its intelligent human meaning and intelligent human fulfillment. Which is the case because the Consensual Matrix already engulfs a multitude of higher beings everywhere, gifted with higher abilities of up to the sixth developmental level, and through them, the Consensual Matrix is capable to identify all its lines of causality future and past wherever it spreads, and therefore it is capable to see and change everything well ahead, remaining capable in this manner to adjust and therefore detail the entire human timeline minutely here on Earth.

While this is a sixth level ability, used now against the human beings of Earth, which are not even allowed to develop to their third intelligent human level, which are not even allowed to exist as living beings but only consensually, which are not allowed to develop and maintain their own higher abilities, which are not even allowed to interconnect naturally among themselves and with their own higher selves, which are not allowed to form a natural, harmonious, human society, and which are not even allowed to care and tend to each other. The Consensual Matrix is so capable in this entire consensual enterprise on Earth, that everything is done by humans against humans with only humans suffering while taking the blame, because this is how the Consensual Matrix always operates, flawlessly.

Who is to blame? Tens of millions of scientists along with

an entire world accepting their scientific lies diligently, and along with consensual neutrinos, Einstein, big bang, masons, weak force, strong force, Schrodinger, his cat, Hubble, and Oppenheimer. While these are only scapegoats, while we can blame them as much as we want, as others are pulling the strings, and we are blaming the wrong actors. Others make us blame the wrong people and the wrong circumstances, as part of the current indoctrination, for an extra twenty dollars in benefits this month.

Study this indoctrination closely, study the lack of accurate knowledge that it enforces, study the social hierarchies, laws, and authorities, and study your social constraints and social conditions, to find them highly consistent in keeping the world exactly as it is currently, an artificial social environmental matrix meant to keep you in it indefinitely, or at least until the end of the genocide, when you are not part of the social equation anymore, but out of the way altogether, since you are of the wrong genes while the current tyrants decide precisely who should be here in this world and who should not.

As a reference, when the covid pandemic started, Gates was in China, experimenting on eugenics on behalf of the invisible kingdom. More precisely, Gates sold his entire fortune as the richest man in the world, in order to experiment in eugenics, on behalf of the invisible kingdom, in China.

Who exactly wins throughout the entire fight for world supremacy? Just study history, because the smart, the just, the kind, and the strong never win as you see in the movies, but the cruelest always wins, to replace the cruel already in power. While it takes a tremendous cruelty to win, to the point where you have no more feelings and no more life left in you by the end of the fight for world supremacy. While you can find all strategy, trickery, strength, cruelty, and abnegation in the Consensual Matrix, since it includes all social actors stated above. The Consensual Matrix always wins, even if it has to replace everyone in it, Masses, humans, invisible kingdom, and the Brotherhood alike. The Consensual Matrix does so for all its worlds and realities even repeatedly, and it always happens on Earth, just study ancient history and old ideologies to learn the truth, accurate and consensual. Yet the truth is always revised before you even get to it, which is called Revisionism.

As another example, according to astronomers and to the entire scientific consensus, stars orbit too fast around the center of distant galaxies, and the laws of physics do not allow a stable orbit at such high angular velocities. Simultaneously, stars from galaxies closer to us orbit normally, by the same laws. Duality again. Physicists asked themselves immediately what holds stars so tightly together in the orbit of very distant galaxies so they can maintain their orbits at such high velocities.

Notice that the scientists never asked themselves if they could have been wrong, and it should be something else making distant stars redshift excessively as though they were spinning very fast. Because the laws of physics in use consensually are never considered wrong, as Einstein's theory of relativity, to be reconsidered and rechecked for errors since the scientific consensus had already accepted them, but it certainly had to be something else wrong out there in the rest of the universe, but not here where we are. Because the current theoretical laws of physics break down when used for galaxies farther from Earth, so there has to be something wrong with those stars in particular and with the entire universe, but not with the erroneous theoretical laws of physics accepted by the scientific consensus here on Earth. This is what the scientists assumed unanimously by scientific consensus, they invented dark matter immediately, and they kept their jobs with pride. While Einstein's theory of relativity remains flawless and invincible, by consensus, along with the entire big bang, now patched up with dark matter that is capable to hold all stars in the orbit consensually, but not really. Why do they still call it a theory, a speculation, or a supposition, if they do not know that it is not real?

Something was taking place there, something concerning only the distant galaxies, and not the galaxies closer to us, or this is what scientists said, so they invented dark matter and then dark energy. However, at a closer study, even invented dark matter or dark energy cannot balance the equations of all natural laws of the universe, the way neutrinos and the entire lepton family of particles cannot balance the decay of elementary particles, because you cannot substitute the natural law of relativity in any manner, by patching it locally whenever needed and for as many times as needed. Similar phenomena should occur similarly everywhere in the universe, further and closer to us, with no place being privileged in the world, because the natural laws of the universe are generally accurate, cannot replace each other, and cannot be substituted in any manner. You cannot consider that things happen in one manner closer to us, and in a different manner farther away, since Earth and our galaxy are not privileged in the universe, and we are not the center of the universe, the way it was thought to be the case until not too long ago. Yet scientists consider that we are privileged, so they formed another duality, a consensus, a lie created only to keep the stars of all very distant galaxies in their orbit according to our consensual laws of physics here on Earth, even while revolving at outstanding angular velocities according to their own redshift. Something very strong, very powerful must be there to keep those stars in their orbits, scientists assumed unanimously by scientific consensus, something powerful enough to attract everything and hold it in place, but something that cannot be detected because it is fiat, not there but only consensual, and they named it dark matter. They should have named it consensual dark matter. They had never detected anything since there was nothing to detect, so they called it dark matter. Yet as stated, even presumed dark matter cannot balance equations of central forces, since the assumed gravitational attraction coming from dark matter is not central for all stars and all galaxies, it is not centripetal, and cannot balance centrifugal and centripetal equations.

The word 'dark' comes from 'undetectable.' Why undetectable? Because only in this manner, your theories can remain theoretically true when they are speculated. It happened with the neutrino, because nothing can ever prove them wrong if they are defined as being undetectable even by the same theories. The same thing happened on the Moon, since they never saw stars there, they never took pictures of stars and of the Earth from the Moon, but only of themselves while playing on the Moon. Yet there are no pictures of Earth from anywhere in space, with the exception of nine intriguing pictures taken throughout the decades, and these are not even consistent with one another. These are very old pictures from the early space explorations, where they still assumed that they could fool the world with everything fake. Just find these pictures over the Internet to see for yourself, but hurry up, because they can be removed systematically before you find them. While they are very careful in the current missions of space exploration not to include direct records as pictures and videos, but only tables of numbers and spikes on a graph.

It is similar with the undetectable neutrino and with the entire big bang, with the dark matter, the dark energy, and then it was the same with all debris from the two skyscrapers since they were minutely removed, while the airplanes themselves were never there, but only fiat, consensual. All dark, all consensual, all undetectable, no more, nothing else to see, nothing else to consider. However, as you study closely all consensual fake fiat records still around from all main fake fiat consensual events as all moon explorations, skyscraper demolition, neutrino records, earth pictures from space, boston bombings, big bangs, dark matter, and dark energy, you find them fake by all laws of physics, while coincidentally, all records from the entire fakery of this world are removed systematically from the human knowledge in every manner. While the invisible kingdom is involved directly in all these with the entire Brotherhood serving diligently while faking everything accordingly, in an undetectable invisible fiat manner. However, since almost everybody is in the Brotherhood currently, everybody fakes and hides everything from themselves in a continuous unnecessary stealth manner.

If you remember that entire large room filled with old

computers and cheering people during the fake first moon landing, all those cheering people from the old computers were killed systematically in a matter of months, in order to hide the entire fakery. Later on, they stopped killing people while maintaining secrecy, but they only changed their identity, which is still good, yet it is still bad that they alter the human knowledge and the human history.

There is true physics in places as Area 51, generating advanced technology, envied by all ordinary physicists of science, if they can ever have access to it, and if they can understand it. Are today's physicists merely actors in the world, just as all reporters, politicians, singers, and mass media influencers? Could this be why the knowledge generated by science is not helpful and not necessary anymore, cannot answer our questions, and cannot fulfill our needs for true knowledge?

This is a poor advancement for science, which keeps people ignorant, resulting in social stagnation, while offering an incompatible lifestyle, keeping people ignorant. When will people ever develop? People do not actually develop in the Consensual Matrix, but the Consensual Matrix only develops the new knowledge, laws, ideologies, and jurisdictions consensually, through agreement alone, meant to accommodate the people of the world, or at least all the brands, trademarks, and corporations that they are considered to be. While these never develop, but only stagnate or decay, exactly as ordered from above. It is similar with all invented illnesses, because if it is ordered from above that you or your loved ones have cancer, autism, alzheimer, cholesterol, covid, aids, or ebola, this is exactly what you have, with all medication that you must take and with all medical intervention that you must undergo killing you. As a reference, nuclear radiation kills with certainty, and this is why you die.

Are the upper social classes to blame? The upper classes are merely scapegoats and you can blame them in vain, the way people blame politicians in vain. It seems that there is something more important, something better organized, better implemented in the world, something consistent, coming from above and beyond our civilization. Same circumstances tend to repeat every time, and they keep appearing everywhere we look. It is a pattern of lies and enforcement implemented across the world and throughout history, forming scientific ideologies but not genuine social domains.

Let us study all available models of the universe. Let us understand what the circumstances were for each one, who created them, who the social actors were, to whom they were addressed, and what had kept people's beliefs steady and in place along the centuries.

5 COSMOLOGY, COSMOGONY, AND THE EVERLASTING BROTHERHOOD

People have always been interested in knowing more about the universe. Cosmology is the science studying the universe as a whole, along with its shape, components, birth or creation, and development. Cosmology is closely related to physics, astronomy, and astrophysics, while in contrast, cosmogony relates to myths of the birth or creation of the universe. Cosmogony personifies the universe or parts of the universe in order to offer a more appropriate, a more entertaining interpretation of the world as a whole.

For example, in the myth of creation from the history of a specific Asian nation, the main deity ejaculated one day to form the Earth, with all the people, the Sun, the Moon and all stars included. Spontaneous ejaculation.

You might find this cosmogonical model at least entertaining, yet at that time, everybody accepted it. It is important to understand it, because it relates to living mythological characters, since cosmogony stands entirely outside the Consensual Matrix, being part of Life, along with the people promoting it. Study your own ideologies now, if you have any, and if the Deity is alive and intelligent in any of them, then you and your ideologies stand outside the Consensual Matrix, which might be extraordinary. Because if your deity is void of life, or if it is only an agreement among people, a concept altogether, or a theory, a formula, an utopic way of existence, a thought, an idea, a virtue, or a symbol, then its entire world is abstract or even consensual in nature with you in it, integral part of the Consensual Matrix. While by accepting the Consensual Matrix, you are against Life, Intelligence, reality, and the living comprehensive Interconnectivity of all living beings. This is how you accept to be consensual yourself, part of the Consensual Matrix, as is your own name written in uppercase letters, which is considered to be you. You are entirely in the Consensual Matrix as a brand, and the consensual model of the human reality applies to you.

In another mythical and religious old ideology, first there was a supreme golden egg, out of which the entire universe was born, exactly the way it is today. Study the dates, to see how this is part of an entire golden age, very distant in the past, consistent with a living, natural, ancient human civilization. While currently, the entire concept of a natural society and civilization is seems primitive in nature, deliberately, since this is the stereotype, influencing in this manner your own natural attitudes and stereotypes against the living real world. This is how you believe that intelligent human communal living is primitive tribal life, related to primitive people as Lucy and the hominids, while people live in very large families even currently, at the top of the human society. It is normal for them and they prosper accordingly, replacing the world while taking you out of their way, probably because you persist to believe that you should live your life disconnected, in a very small family or alone altogether, as part of your own ideology called egoism, the ideology of the ego, exactly as they indoctrinate you along with the rest of the world. The hidden ideology called egoism stops you to see the truth, that you were supposed to live life in a comprehensive intelligent human family spanning the world, since this is the only intelligent human environment that humans should always have by nature, because humans are intelligent and harmonious by nature. While if you are similarly determined not to have children and to satisfy only the multitude of artificial needs and meanings including addictions and servitude while neglecting your intelligent human needs and meanings as everybody else does, then this is the consensual human reality, always apart from the living real human reality, and you fit right in.

The Holy Bible has the Genesis, the book recording the very beginning of the world, with the Deity creating the universe in seven days, while you already know all details. Sumerian ancient tablets describe a similar Genesis, with deities creating the world in seven tablets. Similar Sumerian tablets state all laws and statutes meant to form an entire developed civilization, which is very similar to the current human civilization, while its old ideologies match the current ones, coincidentally matching the entire Consensual Matrix. This might be relevant or not, but the Sumerians go as far back as records can access, and you cannot know too much of how the world was before them, alive or consensual. While the current civilization addresses only laws, jurisdictions, brotherhoods, ideologies, military, business, and all similar agreements, while this is the entire consensual human reality.

There is so much consensus in the world, that you cannot identify and understand the living real human reality anymore, while as you notice, this is the case ever since the Sumerians and long before, ever since the dark ages started. While through cosmogony and through very old records, we still find traces of very old golden human ages, before the dark ages started. While in our entire study of the human reality, we are interested in accurate intelligent human knowledge, because if we want fakery, we already have it in the current consensual human knowledge, and we know it well from the current science, current education, and current media.

The universe is born sometimes from a Mother Deity while other times it comes from the tears, eggs, or sperm of other deities. Being built, born, cried out, created, invented, imagined, or ejaculated, our universe appeared or came to life to become what we know today and what we are trying to understand today. All distinct ways of birth and creation have satisfied consensually the needs for abstract knowledge of all people of Earth from one civilization to another, but which one is accurately true? The story of the birth or creation of the universe varies from nation to nation and from time to time, yet it has always remained consistent with the religion in practice at that time in one hand, and with the demands coming from the leading authorities on the other.

We might find these myths interesting and entertaining, but it was different at the time when they were kept in place. You had to stick to the general beliefs, since if you diverged only slightly, you risked losing your entourage, your family, and even your life, which remained the case throughout all dark ages. You lost your life only for the world to be a golden egg, and therefore a golden egg it remained. What might have motivated people's need for knowledge in the past was the fear of persecution if you were caught believing otherwise, mainly the need for security and the need to avoid death, pain, and oppression from your authorities. Whatever it was, there are still stereotypes from those old times, hidden deeply in science and in our model of the universe. Let us identify them, now that the knowledge of science, history, religion, and mythology is widely available.

Where are these stereotypes hidden within science? What is the true behavior of science? Science seems very rigid and very ignorant, since science is just a consensual social tool, similar to religion. It is an ideological tool, and it is in the hands of the same team of people, being truly divine or only human and local, the Elite. We can easily notice this, since it seems that the ones in control of science are not scientists at all, they do not understand science, and they have nothing in common with science. Because how can you ever fake an entire Moon exploration, by using the laws of physics here on Earth mixed with their own beliefs about the Moon?

Authorities seem to apply a rigid timetable to our

civilization, making sure that it is well set in place and it remains consistent in every manner. Consequently, science does not accept different knowledge, if it is not part of its agenda, and if the scientists in case are not pertinent, well established, and already on the preapproved white reference lists of the scientific consensus, regardless if these know real science or not.

As you notice, consensual science is backwards, similar to all dogma and all ideologies. The consensual science first decides what direction should follow, and then it decides what discovery should be made next, according to its interests. Secondly, science looks around to find all the existing work in that field, to see what it can use, according to its predefined agenda. Thirdly, the consensual science assigns to a specific scientist, someone it trusts well, the task of performing research in that field, transferring to his name all results already found by others. That specific scientist will become rich and famous, will never run out of research ideas throughout his career, his entire work will be automatically accepted by all scientists with everybody hurrying to cite him as reference, and his funding will be unlimited, the envy of all his colleagues, while this seems to be a general rule for a very long time.

We find these privileged scientists everywhere, as Einstein, Hubble, Newton, Galileo, Descartes, Copernicus, and Eratosthenes. You must be chosen, in order to be considered in the world of science, if not, you remain obscure indefinitely and even banished, regardless of the importance of your ideas and the quality of your research. If you are from the Masses or from the Lower Brotherhood, and your research ideas and good results happen to be in the interest of science, matching perfectly its predefined agenda, then your work is transferred to someone else's name, someone who already is or who quickly becomes very famous on your behalf, while you remain obscure and shadowbanned for the rest of your life. You can also run insane and die, if you persist to state that you discovered everything. Yet it is not done to ruin you specifically, but it is done on behalf of the entire Consensual

Matrix, as it is the case with everything in the world.

Is it done in this manner to make those people rich and famous? Is it because they have a higher social status, because they deserve to be wiser and more worthy than you? Yes, but the main reason is because these lucky, chosen, highly pertinent scientists are also highly obedient and highly trusted by science, by the global scientific consensus, and by those above in the Consensual Matrix, trusted to perform the important research work exactly in the manner that they are ordered from above, to obtain the exact results ordered from above, and to stay away and even to disclaim possible secondary results, consequences, and future developments that are forbidden to the Masses and to the Brotherhood. The Elite still use all accurate, advanced scientific knowledge casually, yet the Elite are not allowed to publish it, and therefore they use the accurate, highly advanced science only in private or in the hiding, and this is why the people of the wider world encounter hominids, ignorance, tyranny, and terror while studying Earth, with the same ignorance, tyranny, and terror in their higher worlds, since these are mostly in the Consensual Matrix. While the Consensual Matrix maintains its worlds and realities in a continuous dynastic medieval dark age, for an enhanced control, for an assured disconnection, and for a more predictable and more obedient social interaction. Yet since humanity had known golden ages in the distant past, the human intelligences persist to demand them through all your developmental needs and meanings that you always receive, while in this manner, you maintain a milder human environment all around, not as dreadful as you find it throughout the Consensual Matrix.

Yet the current science is controlled tightly through the Consensual Matrix, in order to maintain the current dark age stable indefinitely, never allowing the golden human ages to return. Because if left freely, all humans fulfill their intelligent developmental needs, while uplifting themselves, the human knowledge, the human society, and the entire world.

Therefore, nothing happens randomly in the current

consensual science according to scientists' free ideas, high inspiration, creativity, or talent as you see in the movies or as you learn in school, since everything is highly controlled, preordered, and well monitored. On a smaller scale, we see this pattern of controlled creativity in art, writing, fashion, movies, and music, and it happens continuously. The Masses remain ignorant of this entire scheme, always watching obediently the movies, and always memorizing science word for word the night before the exam.

Two thousand years ago, Earth was in the Age of Aries, while art and science were highly treasured in the developed cities and in the developed nations of those old times, so important, that they were part of the everyday life. Aristo-cracy means an entire world governed by intelligence, since aristo means intelligence or intelligent. Aristotle, Aristophanes, aristocracy, and the entire age of Aries, since it seems to have been an entire intelligent human golden age in the past, yet long before Plato, Aristotle, Eratosthenes, and Aristophanes, because these refer to even more advanced golden human ages in their own distant past, while always basing their entire research on them. Yet since tyrants destroyed systematically all records of all past golden human ages, they are not in the human knowledge anymore, leaving behind only the consensual human knowledge, which is mostly erroneous and dogmatic deliberately.

Study the current mainstream to see it for yourself, because once you consider social hierarchies and social classes as the highest achievements of the human civilization, there is something wrong in the current human reality. While if you still cannot see it, hierarchy and social classes divide the current human society while making everybody discriminant for all exploitive reasons, always making possible and always enhancing exploitation in life and in the world, considered the highest achievement of the human society and human civilization, and therefore there is something wrong in the current human reality. However, since humanity is always in this manner one dark age after another, it is already common living existence, and you cannot notice it anymore, while believing stereotypically alongside everybody else that the human society should always be divided hierarchically into social layers and social classes. The consensual versus the real, making all discrimination and exploitation possible.

Later on, Alexander the Great conquered Egypt, along with most of the known world of that time. He gave his name to one of the cities on the coast of the Mediterranean Sea, Alexandria, yet he gave his name similarly to dozens of cities throughout the old world, since this is common to all tyrants. Because how can you ever conquer other nations just because you can and just because you are the greatest in the world, as Alexander did, if you are not a tyrant? He did the same with the famous Alexandria, and he assigned one of his generals to rule Egypt as pharaoh.

This example is not about Alexander the Great, since that was only a tyrant, but about Eratosthenes, a famous Greek scientist of that time, who moved to Egypt in order to study the impressive writing, art, and science of Alexandria itself. Eratosthenes might remain unknown to you currently, since only Aristotle, Pythagoras, Plato and Archimedes are more known from those old times and old places.

Eratosthenes calculated the radius of Earth twenty-four hundred years ago. He had also invented the leap day, calculated the tilt in the Earth's orbit, the distance to the Sun, he made the first map of the spherical Earth including on it all meridians and all parallels, and he did everything on his own, incredibly, twenty-four hundred years ago, long before people believed that the Earth was flat and the universe revolved around it once a day. Which is rather suspicious, to invent and discover all these, exactly in Alexandria, the famous city of art and science, which hosted with pride the largest library in the world, that still held all art and all knowledge from the old golden human ages of the distant past. Very suspicious, yet still useful.

The question is not only how and why Eratosthenes was capable to create a more accurate model of the world, but why we encounter outstanding people as him in one century or another, all throughout history. What makes them special? We notice that it is impossible for you to be born randomly, have a revolutionary scientific idea, publish it, and eureka, you invent the best sausage machine ever, capable to remove famine from the world indefinitely while changing the world for the better, and while having all fame and all recognition indefinitely.

This happens only in books and movies, while in reality, the world will never allow you to disturb it from its rigid consensual harmful equilibrium, regardless of how disabled, tyrannical, exploitive, and incompatible it is one dark age after another. Yet we find these outstanding people throughout history as Pythagoras, Copernicus, Leonardo da Vinci, Giordano Bruno, Galileo Galilei, Einstein, and Eratosthenes, who seem to have advanced science more than the others, but how exactly did they do everything?

Something had favored them, there and then, against billions of others alike. Was it intelligence, consciousness, perspicacity, beauty, rapidity in reasoning, physical strength, wealth, influence, or social status? Everybody has these, even currently, and it never makes people too famous to survive the centuries and millennia. As a reference, among hundreds of pharaohs, kings, and emperors of all times, you know only of Plato, who was a philosopher, a scientist, a historian, a teacher, and an author, but not a king or a pharaoh.

How did these people succeed? They had unlimited access to previous knowledge on one side, and on the other side, they enjoyed the unlimited privilege of high connections among the very influential Brotherhood of those times. Because they were in the Brotherhood themselves, while the Brotherhood chose them to be famous scientists from among the others, which is also the case currently.

The Brotherhood gives you the fame, status, wealth, and recognition that you desire, if you serve it well, and if you do something remarkable according to its agenda. More precisely, the Brotherhood does everything, while only using your name. Otherwise, if you happen to do all these great things alone, the

Brotherhood erases you from existence altogether, even if you are a king, an emperor, or a pharaoh. It is important to see why all the famous people of the past are currently praised, as Einstein, Napoleon, Newton, Cleopatra, Edison, and Tutankhamun, because as we always notice, intelligence, strength, kindness, and outstanding knowledge are never enough, but you must be in the Brotherhood to have them, and through it, you must be in the Consensual Matrix. Yet this is the case only throughout the dark ages, when the Consensual Matrix engulfs Earth, because throughout the golden human ages, humanity is free, intelligent, and always successful, while everybody capable is recognized accordingly.

This is the case because the Consensual Matrix spanning most of the wider world orders the Brotherhood directly everything that must be done in this world, in the human reality. While as you notice, this is only the consensual human reality, yet it is enforced to replace rigidly the actual human reality, which is this world.

While as you notice, we tend to study more physics and the Consensual Matrix throughout our model of the human reality, since physics studies directly this objective reality, while the Consensual Matrix orders everything in this world, including all achievements in physics, and all physicists themselves. This might seem impossible, but it can always be done consensually.

Because throughout dark ages, this entire world belongs to the Consensual Matrix, legally, with all the necessary certificates and licenses to prove it valid, lawful, and legitimate, since the Consensual Matrix itself writes the law here on Earth and in most of the wider world. EARTH incorporated. Yet the Consensual Matrix owns only the consensual part of the world, as the Consensual Matrix states, yet the scientific consensus of all times along with its entire list of theories, discoveries, and inventions are consensually accepted through consensus and therefore they are already part of the Consensual Matrix, while defining in this manner everything real and everything consensual as it wants. The Consensual Matrix does the same to zillions of worlds and realities simultaneously, and therefore it has a very well defined agenda stating exactly what theories, discoveries, and inventions must be made, how, why, where, and when. While with the Brotherhood and the Elite of this world implementing everything diligently, they choose your name in science to make everything possible, exactly as ordered through the Consensual Matrix, exactly as it was the case with Einstein, Leonardo, Eratosthenes, Edison, Newton, Thompson, Hubble, Heisenberg, and Pauli.

It might seem incredible, but all Brotherhood advantages had allowed them to create and publish scientific works that were outstanding, revolutionary, yet highly banned by the authorities of those times. How interesting, because if you are stubborn and insist to publish your scientific work even when you go against science, religion and authorities, then your friends in high places can support you as best as they can for as far as their interest in you lasts, and for as far as you can provide to them.

While when you study these great historical personalities, you find them involved exactly in the change of those specific world orders, allowing, helping, and therefore participating in this change of world orders themselves. Therefore, if you were capable enough to help change an entire world order, mostly according to the specific interests of the Consensual Matrix, you were rewarded with everything that you desired, including eternal fame, as it was the case with Einstein and Tutankhamun, and with everyone else stated above. You were rewarded with unimaginable wealth and influence, as it is the case with Oppenheimer, who even received an entire nation as a gift, and they might still have it or they had already lost it. South Africa.

With what consequences for the world? That specific nation suffered of apartheid continuously under Oppenheimer, and it might still suffer today. Similarly, Egypt returned entirely to the dark ages through Tutankhamun alone, while the entire science ended up derailed by Einstein, dumbing down an entire civilization. Millions died in atomic warfare through those particle physicists mentioned previously, while an entire invisible nation, the invisible kingdom, took over the entire world, to exploit it and even to eradicate it at will, through all famous scientists, singers, actors, investors, billionaires, and presidents. What souls exactly want to come in the invisible kingdom and in its harmful ideology? Not the developed, but only the tyrannical undeveloped ones, which is also the case with the dictators of the East.

Do you see the discrepancy between the Brotherhood and the authorities? They are not the same throughout the ages. Who is really in control? The Brotherhood, but they seem to be in control only of what they can manage to control, since this is how they control only the Masses on behalf of those above. Yet with most of the masses already gone, they control only themselves, in a continuously exploitive puppet show, while exploiting themselves and while praising themselves simultaneously.

How would you like to be born in Egypt in those old times, making all these important discoveries yourself, since it seems rather easy to draw some meridians on a map? Yet could you make all these discoveries yourself? Let us see.

Eratosthenes was not actually born in Egypt, but he came to Egypt from Greece, to become none other than the mighty Chief Librarian of the mighty Library of Alexandria, containing all the knowledge in the world. If you came to Egypt in those old times to manage farmlands or shoe stores, you still had some fame. If you came to manage the entire Treasury of Egypt instead, you had your profit, and you became very rich. Yet if you came to manage the greatest Library of Alexandria, you managed to steal all the knowledge yourself, placing everything systematically in your name. This worked well, mostly since Alexandra underwent a drastic change of regimes, while you were Greek, sent to Alexandria by Alexander the Great himself to manage the Library of Alexandria, making you famous in this manner, the greatest scientist of all times, Eratosthenes himself.

The Library of Alexandria was not exactly similar to your local public library, since it is rumored that it had held knowledge inherited directly from the greatest golden human age, Atlantis itself, everything that Alexander the Great gathered from the past age of glory to place there, along with everything that was already there, a very great amount of knowledge, currently mostly gone.

Now you can understand better Eratosthenes' wonderful discoveries in science, along with his high status within the Brotherhood, because everything achieved scientifically in that old golden age was transferred in the name of Eratosthenes, the great Chief Librarian of the Library of Alexandria itself, the greatest library of all times, holding all knowledge of the previous age of Earth and more. Since this is why it was the greatest library, because it contained records of higher knowledge from the previous ages of Earth, the golden human ages, currently erased systematically from the human knowledge.

How convenient for Eratosthenes, and how convenient for the Brotherhood and the Consensual Matrix just as well. Because in this manner, they had managed to erase the true, real, living human past of Earth, transitioning everything from an intelligent human civilization to the current consensual civilization. Eratosthenes kept his mouth shut about the previous civilizations of Earth, he erased its traces systematically, with some scientific achievements of the past transferred to the name of Eratosthenes himself.

Yet there were different times then, because highly intelligent people were still respected, treasured, and acclaimed by everybody, since they even made you a leader if you were smart. Aristocracy. The word 'Aristocracy' means intelligent leadership, intelligent authority, or an entire regime formed of the smartest people of that time.

If this was not enough, Eratosthenes ordered everybody to bring to his library all their private manuscripts, everything written, in order to have everything duplicated at the library and then they could have it back, while in this manner, he expanded significantly what was already the largest and the most famous library in the world. Eratosthenes exchanged ideas with his closest friend Archimedes, and rubbed shoulders with the pharaoh himself, while he even tutored his children. Yet Aristotle also rubbed shoulders with the pharaoh and tutored his children, since all very smart people were very famous and very praised in the Age of Aries. How did Archimedes know all that knowledge that you learn in school today? Only Archimedes and Eratosthenes knew it, along with the rest of the Brotherhood.

The current science states that Eratosthenes calculated the circumference of Earth by measuring the shadows casted by two equal rods at noon on the summer solstice, having one rod in Alexandria and the other one in the Elephantine Island. It is said that he came close to the value in use today for the radius of Earth, yet his mathematical errors had exceeded by far today's expectations. Study his experiment closely, to see how his poor approximations downgrade the very ingenious idea of measuring the shadows of the two rods in order to find the circumference of Earth. He had wrongly estimated that both Alexandria and the Elephantine Island were on the same meridian, probably giving him a small error. Furthermore, he determined the distance between the two rods not through an actual measured distance, but from the time that it usually took camels to travel from Alexandria to the Elephantine island, a very inaccurate method. Camels actually traveling to the Elephantine Island, since it seems that camels went there, and we will see how.

After all inadequate estimates, Eratosthenes averaged everything to only one or two significant figures, which could have resulted in errors as high as ten percent. Eratosthenes could have found the idea of measuring the shadows of the two rods along with the necessary geometric formula in one of his priceless papyri from the library of Alexandria, he could have invented it himself, or it might have been revealed to him by Archimedes, since this ingenious idea seems conspicuously consistent with Archimedes' thoughts, interests, and work.

Whatever he did, his error is of less than sixteen percent if he had used the Greek system of units, or of only one percent if he used Egyptian system units. Which is remarkable, if it was actually his work, and not already part of the mighty Library of Alexandria, where he was the Chief Librarian, and where they always used Egyptian units. Yet how exactly were those camels travelling to the Elephantine Island like submarines?

All circumstances are contorted systematically in order to allow the entire transition of very important scientific knowledge to the name of Eratosthenes. The Elephantine Island was a larger hill during the ice age of Earth, long before Eratosthenes, when the people of the Lower Egypt discovered everything, while they lived life in an actual golden age. The water level was one hundred meters lower then, and you could travel to the Elephantine island by camel, thousands of years before Eratosthenes, while there might have still been elephants in the area, when this particular experiment had been done. While as you study all details closely, you find everything consistent.

Since it is possible that this ingenious experiment is the work done by the scientists of Lower Egypt further in the past. There is a difference between the Lower Egypt and Higher Egypt, obscured by the current science. The Sphinx marked the boundary between these two powerful nations of the distant past, the Lower Egypt and the Higher Egypt. These were separate nations, having separate cultures and separate genetic lines. All knowledge from the library of Alexandria came from the Lower Egypt, not from the Higher Egypt. Because even further in the past, during the past ice age, the Lower Egypt was part of the mighty Atlantis itself, while the Higher Egypt was not. This entire knowledge is currently obscured by the current consensual science, since this is how the Consensual Matrix wants it.

The Consensual Matrix erases persistently the entire actual human reality, replacing it with the consensual human reality, while in order to do so, it has to erase all the golden ages of Earth spent in the genuine human reality, with Atlantis included, taking everything intelligent out, while leaving in the human history only the consensual human reality of the

Consensual Matrix, only the current Egypt, which is the Higher Egypt of the old times, along with Greece, Italy, and Mesopotamia that you know well. Yet since the Lower Egypt was part of Atlantis while the Higher Egypt was not, the Consensual Matrix erases the Lower Egypt itself from the human history along with its entire successfully developed civilization, but not the Higher Egypt, currently called Egypt.

Why is one higher and one lower? Higher Egypt and Lower Egypt. One Egypt is actually lower in altitude than the other, only one hundred meters, yet throughout the succession of warm ages and ice ages on Earth, the Lower Egypt submerges underwater as it is currently the case during this entire warm age of Earth, while the Higher Egypt remains above water. E-gypt means high gypt, high earth, or high land, while gypt means earth or land. Gyptians or Lower Egypt. The Lower Egypt built the Sphinx and the pyramids, consistent with their own culture, tradition, and civilization. What else do the Brotherhood and the Consensual Matrix erase from the human knowledge? Everything, the entire real human reality, replacing it with the consensual human reality.

What exactly is going on, with submarine camels and islands coming out on land on their own? Everything is related to the fact that the level of the Mediterranean Sea varies by hundreds of meters between the two significant ages of Earth, the ice age and the warm age. This is how the specific niche of the human civilization moves continuously throughout the ages, while maintaining itself on land and not underwater. While in the specific place of the Mediterranean Sea, water levels descend almost two hundred meters into the Mediterranean Basin during the ice ages of Earth, to form the Lower Egypt, above waters. While during the warm ages of Earth, as it is the case currently, you have people living hundreds of meters higher up inland and away from the increase of the water level of the Mediterranean Sea, due to the melting ice, leaving above waters only the Higher Egypt, which is Egypt, as it is currently called. The Lower Egypt was part of the Atlantis civilization of the old age, along with the entire

Mediterranean Basin of those times as it was only partially submerged, forming Atlantis.

Atl-antis means the ancient world, or the antiquity as it is currently called. However, Atlantis was the antiquity of our antiquity. Atl-antis means the world before, the world before this world, or the last age. While anti-quity means the world before they died or the world before they left, which is also the world before, or the last age.

Plato lived in our antiquity, while making references to his antiquity, the previous age, Atlantis, which was his antiquity, or the antiquity before antiquity. Eratosthenes also live in our antiquity, while also using knowledge from before his age, from his own antiquity, the antiquity before the antiquity. However, there was a golden human age so successful and so prosperous in the distant past, that it managed to span entire warm ages and ice ages continuously all around the Mediterranean Basin, in South Europe, Middle East, and North Africa, sustaining a very developed human civilization. Plato found it referred to as Atlantis in the very old records, and this is how he described it in his books.

As it was the case in the entire Atlantis, the people of the Lower Egypt of those times were interested in all relevant knowledge of science, civilization, and art. Their national emblem was the papyrus flower, the symbol of intelligence, learning, writing, and studying, while their deity was the Snake Goddess, the symbol of intelligence and spirituality, also depicted by the Sphinx, before it had been remodeled throughout the following dark ages. The Sphinx marked the border between the Higher Egypt and Lower Egypt of the old times. Where else on Earth can you find science, art, and spirituality so close together? In the Ancient India, also contemporary with Atlantis. However, ancient Europe was also very successful and very developed in the distant past throughout entire golden ages, yet all very old European records were continuously erased by all consensual Brotherhoods throughout all following dark ages of Europe.

As a reference, Napoleon remodeled and repainted the

pyramids when he conquered Egypt, rendering everything consistent with the European religious ideology of his time, yet all tyrants erase the human knowledge continuously throughout their reign, leaving behind only the dogmatic, the ideological, and the consensual. Study the current North Korea closely to see it yourself.

Both Alexandria and the Elephantine Island were part of the Lower Egypt when the experiment was made to discover the exact radius of Earth, yet it was a different Alexandria then, several kilometers to the north, currently submerged. While as you study Plato's books closely, you notice how he referred to this sunken city as part of Atlantis, the sunken Alexandria several kilometers north, currently submerged.

In this manner, the old sunken Alexandria is on the same meridian with the Elephantine island, and you could travel from the old Alexandria to the Elephantine hill by land since they were both above water, and they were both part of the Lower Egypt, both having universities and research centers studying Earth along with its radius, longitudes, and latitudes.

Furthermore, you cannot simply place some lines on the map of Earth to call them longitudes and latitudes, but you must do so in a spherical system of coordinates, you must know accurately the radius of Earth, while you must know the entire trigonometry to do so. Coincidentally, they used sixty digits in their system of numbers but not ten as they are currently used, maintaining consistency with the entire current trigonometry, as though the current trigonometry is copied entirely form the distant golden human age, while using sixty digits to define time in hours, minutes, and seconds, the displacement of all space objects in the sky, and the displacement of all places on Earth by using longitudes and latitudes.

This is the work of thousands of scientists while maintaining consistency with an entire continuous golden human age kept on all intelligent human values but not on ideologies and irrelevant knowledge as it is the case throughout all dark ages, however, if Eratosthenes discovered all these on his own while partying in the high society of the old Egypt and while tutoring the children of the pharaoh himself, then this defines exactly tyranny and dark ages, since as it is the case with Einstein, Pauli, and Hubble, this people do not even understand their own statements.

They might have used camels or not to measure the exact distance between Alexandria and the Elephantine hill, yet everything points out to the incredible knowledge of those old times before Antiquity, during the ice age, when the Lower Egypt was above water, when the Mediterranean water levels were hundreds of meters lower than what we currently have, when the human civilization surrounding the Mediterranean Basin and probably spanning the world was in its golden age, contemporary with the Atlantis civilization or part of Atlantis itself. Contemporary with the ancient Indian civilization just as well, since when you study the Lower Egyptians and the Indians closely, you find them similar in many details.

We notice how Alexander the Great conquered Egypt, gathered all the necessary knowledge from the past age of Earth, sent his own general as pharaoh there, and then he sent his own chief librarian to manage all knowledge, Eratosthenes. This is how they stole and divided the knowledge as they pleased, and now you learn only about Aristotle, Pythagoras, Eratosthenes, Thales, Plato, Archimedes, and Euclid, while they used the knowledge of the last golden human age in everything that they researched.

How did they measure the radius of Earth with a precision of one percent, and with all longitudes and latitudes included exactly as we currently use, thousands of years before humanity even discovered that the Earth is not flat? What are they hiding? Because when you hide the real human reality while switching to dark ages, you destroy the real human knowledge, the real human meaning, the real human civilization, and humanity altogether, in order to render everybody ignorant and undeveloped for exploitive reasons. This is the actual transition from the living intelligent real human reality to the consensual human reality, as it takes place every time you transition from golden human ages to dark ages. However, you still have to keep some knowledge in order to allow your servants to serve you at decent level, by having some knowledge to measure time and distances.

All circumstances changed after Eratosthenes, during the following age of Earth, the Age of Pisces. Religion controlled science and education entirely, shaping and reshaping the model of the universe the way it pleased. Today's religions are very mild, friendly, and excessively helpful, while in the past, religion was the perfect tool in the hands of authorities controlling the masses. This is not entirely true, since it seems that religions controlled authorities themselves. While powerful nations and empires managed to conquer religions only by force and invasion, while even then, they only confined religions, temporary, the way the Romans never dared to invade Vatican, but only kept the Pope confined there for decades.

By studying all paradigms throughout history and by studying what had caused each one to flip, we notice a striking resemblance to the modes of living of an organism, when it switches you from one mode of life to another allowing you to cope with the environment in order to stay alive and fulfill all needs in the most efficient manner. If we consider similar modes of life for entire civilizations, then these modes of society and modes of civilization are predefined and are stored in the common genetic memory. In this case, the flip of the paradigms or the change of modes of civilization happen because of intrinsic reasons. Higher beings and powerful elites do not have to design a new mode of civilization from scratch to match their greed and conquering agenda, because they can trigger society to switch itself to a different mode of society, one that matches the most their consensual interests. Society will not only flip its mode that way, but will cause the entire civilization to flip its paradigm or its mode of existence, either naturally through natural environmental constraints, or consensually in the favor of the Elite and the entire Consensual Matrix.

The Consensual Matrix knows well how to control all modes of civilization throughout the wider world, and it gives the necessary orders here on Earth and the Brotherhood from Earth works overtime to make it happen, destroying entire golden ages of Earth in the process and making the world and the human reality as consensual as it can ever be, with pride, virtue, and heroism, since the Brotherhood takes pride in its entire consensual work here on Earth. While this is the basic consensual ideological servitude developmental level.

This is what the De Medici family did half a millennium ago during Renaissance, as they were in the Brotherhood and they fulfilled their orders well. This is what the Rothschild family did a few centuries ago while triggering the Industrial Age, deliberately or not, by contorting the genuine Brotherhood of those times. While on a smaller scale, this is what the Rockefeller family did in the Americas throughout the past century, to trigger United States.

The modes of civilization are more distinct when studied on a larger scale, with their pattern following closely the zodiac or the circle of animals: the bull, the ram, the virgin, the twins, and the lion, all the animals of Earth, from the word 'zoo,' signifying a gathering or enumeration of animals. Yet who exactly are these higher beings controlling Earth in such a systematic manner and for such a long time through the Consensual Matrix, while considering the people of Earth animals? There are many higher beings controlling and contorting the Consensual Matrix here on Earth. Search the Internet for the name of the deity controlling the current invisible kingdom. It starts with letter B, yet it is considered a demon by many of the higher beings roaming around Earth.

The zodiac from the Western Civilization is an ancient invention associated to the major cultures from around the Mediterranean Sea. The basin of the Mediterranean Sea is associated with a highly advanced civilization, now deceased, submerged, dating from twenty thousands to twelve thousands years ago, called Atlantis by Plato, or Atlantida by the ancient Greeks. The zodiac is not associated to Atlantis, since the zodiac has a cycle of 25920 years, matching the movement of precession of Earth, and it could have been in use during all previous succeeding civilizations thriving between Europe and Africa, with Atlantis included.

Why did they need a zodiac? Why did they need pyramids? Because the place where these civilizations appear and thrive is under water for half of this cycle of 25920 years, and you need to know exactly when conditions are proper for that civilization to appear, and then when it starts to submerge and therefore end. The zodiac tells you not only when the land is above and below water, but it tells you the stage of that civilization and even the climate along the millennia.

Who uses the zodiac? Who needs it? Normal people measure their lives in years and decades. Dynasties, countries and empires measure their existence in centuries, while civilizations span for a few thousand years at most. It seems that only higher beings or deities have made use of the zodiac, and probably still do. It is also possible that those temporary, cyclical civilizations from the Mediterranean area achieve such an outstanding progress in one time or another, forming a model and an understanding of the world so accurate, that they understand the Earth's ice ages and can therefore predict them along with the sea levels throughout the millennia. A third possibility is that people of these civilizations advance to such a high spiritual level, that they have direct access to past and future events, they see them, predict them, and the zodiac is their way of transmitting their knowledge to future civilizations, warning them of the cyclical sinking and reemerging of the world, while you already know the story with the deluge.

Because there is more to long term zodiacs and calendars, since through higher abilities, you are capable to see directly past and future events, and you need a very large time scale to be able to state with precision when and where they take place.

There is a relatively modern zodiac system in Vatican. The tall column from the square, along with the specific pattern of crossing lines on the ground allow precise measurements of the tilting of Earth, and implicitly, they allow the measurement of the slow movement of precession throughout centuries and millennia, which coincides with the succession of the houses of the zodiac throughout the 25920 years cycle.

Why having to be so precise? All information that you can find in this subject is that someone or something comes here on Earth at the beginning of a specific zodiacal age or at the beginning of all ages of the zodiac, and the Vatican is highly interested in the precise year and day when this happens. While with a giant pinecone in Vatican in the same place, which is the symbol of the pineal gland and therefore of the human spirituality, with the Snake Goddess as the main deity of the Lower Egypt, also a symbol of spirituality, and with a significant amount of ancient records of higher knowledge stored in the Vatican underground, there is something going on. This entire knowledge is hidden and it is used currently by the Elite and by the Higher Brotherhood, yet you can still read the old mythological, spiritual, and religious records of Ancient India to find the truth. Hurry up to read them, while they are still around.

The truth is that once developed at your intelligent human level, you have your abilities developed, and these can be as high as your own genetic line has managed to develop them throughout generations, races, species, kingdoms, and forms of life. In this manner, though your higher abilities, you are capable to overrule the Consensual Matrix even alone, you are certainly a liability for the Consensual Matrix while developed, and more importantly, the Consensual Matrix knows about you well in advance.

This is why currently, there is no threat for the Consensual Matrix on Earth coming from the people of Earth and from their higher selves, since it made sure well in advance to clear up Earth of all possible developed threats now and for the eons to come, since science is capable to remove systematically all developed, free genetic lines as it can see their behavior well in advance, leaving behind only the obedient genetic lines and all their highly developed abilities, to serve the Consensual

Matrix indefinitely.

While with the Lower Brotherhood replacing the Masses altogether, you have continuous dark ages on Earth, with everybody serving piously within a minutely choreographed and highly obedient everlasting Brotherhood. You are with the Consensual Matrix or you get out of the way, and you can already see the souls making their choice well, while all ideologies state it clearly, since all current ideologies are the ideologies of the Consensual Matrix.

This is only the consensual agenda, yet it does not have to be in this manner, since as long as you are born here, you have the right to live your life as freely and as naturally as you please, exactly as your soul desires. Therefore, you can turn this place into an intelligent living world if you please, free of the entire Consensual Matrix, while you can do so alone or alongside everybody else.

On a global level, the changes of the modes of civilization happen in longer, equal intervals of time, not according to the level and rate of development of the human civilizations, but according to cyclical time measured in thousands of years that match perfectly the zodiac and the precession of Earth. Major natural cataclysms as ice ages, floods, and continuous droughts are relatively cyclical on Earth, and are associated with the precession of Earth, among other factors, triggering in this manner the death and rebirth of entire civilizations. This is Phoenix from the Brotherhood ideology.

Not all ages of Earth are associated with major cataclysms, yet civilizations still seem to change their modes of existence eventually. Someone or something wants it in this manner, or it is only the longer-term change in the environment, and all civilizations are capable to switch their modes accordingly, just the way human beings are capable to shift subconsciously from one mode of life to another, depending on the specific longer-term needs that they have to fulfill.

The precession of Earth makes a full cycle or 360 degrees in about 25920 years, called one world. Thirty degrees of precession is equivalent to 2160 years, which is called one age.

Thirty degrees on the celestial sky are as wide as a full constellation. There are twelve intervals of thirty degrees in the entire 360 degrees cycle, and coincidentally, there are already twelve constellations in the sky defining the ages of Earth, the signs of the zodiac.

How do certain constellations define the ages of Earth? Watch the Sun at sunrise at the spring equinox. We know that not only the Sun moves across the sky during the day, but all stars do, while accompanying the Sun, yet we cannot see the stars, since the Sun outshines them. However, at sunrise, you can still see stars in the sky, if there is no pollution. The stars accompanying closely the Sun on spring equinox are part of one of the twelve constellations associated to the zodiac, and they define the age of Earth. The Sun is in one of these twelve houses during the 25920 years precession at Easter or spring equinox, and it changes houses or constellations every 2160 years, or every 30 degrees, moving in this manner slowly from one constellation to another.

The giant sundial from Vatican can define precisely the ages of Earth. However, the true ages of Earth are not defined using calendars, but by simple observation of the sky, by the house or constellation that clearly accommodates the Sun in the sky. Some constellations are wider, and consequently, these ages last longer. While other constellations are narrower, making their age shorter.

It is said that for each age of Earth, there is a specific deity ruling the word. This might be true or false, depending on what you believe, yet this seems consistent to most religions. During the last age, the Age of Aries, the deity was associated to a ram, a goat, Pan, or Aries, a deity with horns and hoofs. We still see this personified goat in the old art and in the old religions. He plays his bone flute, he sings, dances, and performs magic and alchemy, which were part of the science of that time. That goat, ram, Aries, or Pan was associated with all arts, sciences, religions, architecture, and the spiritual development of the people of that time. It was the age when scientists as Pythagoras believed in a spherical Earth revolving around the Sun, and they had all the necessary geometry and scientific concepts to prove it.

This is not only a simple belief, but it was the age of Earth when Eratosthenes had 'calculated' the circumference of Earth, the distance to the Sun, he sat in place the modern calendar, and he created the map of Earth with meridians and parallels, all by himself.

While as you notice, it is not as science states currently, that people evolve gradually from the old hunter and gatherer primitive ages of Earth to the horse and wagon ages of Earth when everybody assumes that the Earth is flat, and then to our wonderful modern age when we should thank science for all the scientific development that it is capable to offer. Because as we notice, there were ages of Earth so developed and so prosperous, that were called golden ages, with information so precise about this world and about the entire human reality, exceeding what the current science can offer. If we consider that this entire old knowledge from the time of Eratosthenes and Pythagoras is only a small remnant of what was in the past age before them, during the last ice age of Earth, we can certainly wonder what else the current science hides from humanity and why.

Religions change with each age of Earth, and they bring new deities to be venerated, while the old deities become the anti-deities, the devils and demons, as it happened at the end of the Age of Taurus and at the beginning of the Age of Aries, when people were made to believe in the goat, the ram, or Pan, while they were made to fear and despise the bull or Taurus, which became the anti-deity. The same happened in the following age, the Age of Pisces, controlled by Christianity. The goat, Aries, or Pan became the anti-deity, Satan, the devil, or the demon, constantly tempting to steal and pervert people's soul through music, dance, literature, works of art, and scientific knowledge, which are intelligent human abilities and intelligent human achievements.

The word 'Satan' means 'follower of Seth.' Seth was a major deity in the Upper Egypt, Lower Egypt, and Lower

Canaan area, now submerged, representing the wisdom coming with the old age of people, and the old age of nations and civilizations. If you see pictures of pharaohs carrying a specific cane even when they are young while helping them to walk, it is to depict them as being old, wise, and followers of Seth. All good knowledge, habits, and values associated with the Age of Aries became heretic during the following age, the Age of Pisces. Seth and Satan also became a negative deity in the Upper Egypt when his Age was over. 'And I shall be with you until the end of the age.'

We find similar traces throughout the longer history, yet people never assume that the Age of Pisces was but a temporary age on Earth and that it had already ended, as it made room for the next age, the Age of Aquarius. Christianity can continue to control the Age of Aquarius in the Western Civilization, the way it controlled the Age of Pisces, or it can depart, making room for a new deity as the Bibles state, along with a new religion, and new beliefs.

If Christianity departs, be ready to witness new governing methods, new sets of values, and new models of the universe, since a new age will bring new ideologies and a new lifestyle with it. A significant religious character said: 'I will be with you until the end of the world, or until the end of the age, and nothing will be in the new age as it was before.' Do not expect a nice farewell for Christianity, since by the way our society behaves, everything old is bad and must be destroyed when it departs, blamed for all atrocities, while everything new is perfect, it is welcome to come, and it is the new savior. It is similar in politics.

Everything associated with Christianity might become the opposite, the anti-deity during the new ideology. Is this good or bad? All ideologies are consensual in nature, and therefore are of the first servitude level. Again, it is your choice if you want to live your life undeveloped at the first dogmatic level, or developed at the intelligent human level. While undeveloped at the first servitude level, everything remains unchanged from one consensual age of Earth to another, with the same hierarchies, duties, artificial needs, exterminations, and continuous servitude. Only the ideological theme tends to change, and it might be social, national, religious, spiritual, scientific, juridical, addictive, or cultural.

Yet it is just the same servitude, only manifesting in a different form. While even the people throughout hierarchies are the same, since they tend to keep about the same hierarchic spot from one age to another and even from one regime to another, if they are not unfortunate enough to be the ones being the scapegoats. Yet people are capable to change their minds, their sides, and their beliefs overnight, regardless of who is in power and what religion is implemented. The same happens at the beginning of each term in politics, when people turn around to disclaim the very people they once voted for and supported.

Religions last for thousands of years or less. People are not alive that long to see both the beginning and the end of each age of Earth and of each religion, how good they are in the beginning, and then according to authorities, how bad they become in the end, and how all values turn upside down with each new age of Earth and with each new religion set in place. If you see very important people displaying inverted symbols, these depict the fall of a specific age of Earth or world order on Earth, along with their explicit embrace of the new age, stating clearly if they are in the new age, new regime, or new world order or not.

All paradigms are only temporary, as they change simultaneously in science, society, art and religion, being constantly replaced by newer, 'better' paradigms. In the same manner, all revolutions are always temporary, and they always make room for new revolutions, new political parties, new social structures, and new regimes, in a matter of decades, doing the same, which is always the case in an undeveloped world. When you study everything closely, you understand that what is always consistent throughout time is only people's hard work, people's suffering, and people's fight for one 'noble cause' or another, triggering people's dramatic involvement in this highly controlled preprogrammed existence.

All models of the universe change with the modes of existence of each civilization, which seems to match not the actual level of development of each civilization, but it matches an expected level of development of the civilization, the way students' learning levels match not their temporary abilities and achievements, but they match distinct teaching programs and entire curricula implemented with precision in sequential school years.

These sequential changes in the modes of civilization explain the real behavior of science, since the models of the universe in place at one time in history or another have nothing in common with what people thought, imagined, believed, invented, and discovered at that time, but they are related only to what people are expected to know and to how people are expected to behave at one specific time in history or another. This entire enforced pattern of values and beliefs changes drastically. What you were forced to learn, believe, teach, and say during one age and religion, becomes quite the opposite in another age and religion, and you can lose your freedom and life if you keep the old knowledge, beliefs, and models of the universe.

In another example, Copernicus was the first scientist to observe, notice, think, understand, and believe that the universe, including the Sun, does not revolve daily around Earth, but it is quite the opposite, Earth rotates around its own axis, and it revolves around the Sun.

Were the old models of the universe direct outcomes of the general ignorance of the people? Were they direct outcomes of a poor education, or of a simple misunderstanding of the laws of physics? No, but people were obligated to accept the official knowledge even if they thought otherwise, this is what they did, and that was exactly what they thought. If you were there, you probably understood it better, and you were probably the first one to state that Earth is flat, standing right at the center of the universe.

Was the inquisition threatening or torturing you? No, since

how could the Inquisition threaten or torture you along with hundreds of thousands of people? This is only a stereotype, depicting all people oppressed by the wealthy and the powerful, while being forced to work, think, and behave in a specific manner on behalf of the Elite. This is not true, because the Elite of that time, which the royalty, were busy indulging themselves in a beautiful everlasting abundant life, having nothing in common with the Masses. Who controlled the Masses then? The aristocracy and religion did, the way people assumed and still assume currently, but how did they really control the Masses? Did the priests and nobles go in person to oppress all the masses back then as history states? No, the masses controlled the masses, with little intervention from the social classes above. More precisely, part of the masses, the greatest majority, watched and made sure that everybody behaved, believed, and spoke according to the rules. Brothers control Brothers, peers control peers, colleagues control colleagues, coworkers control coworkers, people control people, or citizens control citizens, and they do through the need for social acceptance and social competition, or for avoiding being ignored and excluded. Everyone abides automatically to the norms of the group, entourage, peers, and society, if not, you downgrade socially, you become the last of your group, the scapegoat, the banished, or the oppressed, you are excluded from your entourage, you can lose your family and close friends, you might have to leave town, you might try to seek revenge ending up losing your freedom, or you might even die. Masses oppress masses without hesitation.

Could anyone ever bend the norms and directives coming from very high above? Yes, but it takes sacrifices and an entire network of powerful people to do so, and this happens only when there is a very important agenda at stake. Who are these relatively important people, daring to go against high orders coming from above? They are either a powerful part of the Brotherhood, or they are a significant part of the Brotherhood.

What is their intention? These are exceptional circumstances, they do not happen too often, and they are always associated with the social structure or with the social order. What happens is that, in time, some families become wealthier and more powerful than others, and when this happens, some families are overtaken, and they lose social status. The outcomes are relatively minor, and they are noticed only locally. However, other times, entire factions of families ascend socially, through major wars and terrible crisis, and it rocks the entire society, while bringing in place a new social structure, a new social order, the new world order. While when this new world order is significant, it triggers the change of the paradigms. Everything old becomes a liability and it must be quickly changed, since it always threatens to reinstate the old world order, once again.

Copernicus was exactly in the middle of all these. He worked hard his entire life for it, then he lost his life for it, and he probably never knew it. Let us see what happened.

Copernicus did not only observe the sky one day and had the ingenious idea that everything was a matter of perspective, and therefore the Sun does not revolve around Earth but Earth rotates around its own axis making it look as the Sun, the Moon, the stars, and the planets revolve around Earth, but Copernicus was privileged to have access to very old knowledge, very old books and manuscripts banned by the Church, remnants of the previous, prosperous civilization of antiquity, the Age of Aries, which was basic blasphemy at that time. This helped him understand the world, teaching him about the heliocentric and the geocentric models of the world, teaching him what was true and what was only an illusion.

Was it a coincidence that, before Christianity, the pagan religion was oriented towards the Sun, which gave us light and energy? It made sense to believe that Earth took its energy from the Sun, Earth depended on the Sun, and Earth also revolved around the Sun. Pagans and Arians were oriented towards nature, Mother Earth, and general archetypes associated directly to beauty, war, muses, art, magic, heroism, living knowledge, study, research, and teaching.

What made people of Ancient Greece, Ancient Rome, and

Ancient Egypt believe in a heliocentric model of the world if not their own religion, the religion in use during the previous age of Earth, the Age of the Ram, right before Christianity? Do you see how there is not a simple invention or discovery that Copernicus might have done randomly, since he had access to an entire collection of old banned knowledge, covering philosophy, art, magic, science, religion, history, and society?

At that time, the entire scientific and artistic knowledge was called 'the occult,' which means the hidden. Scientists were also alchemists at that time, and they used science while trying to transmute metals and create gold in order to become rich, or trying to find the elixir of youth and immortality.

The geocentric and heliocentric models of the world were not only a matter of perspective or a matter of taking sides randomly, choosing one model of the world against another, as you chose your political party or your favorite football team. The reason why Christianity chose the geocentric model of the world, which is the model where the entire universe revolves around Earth once a day, is because it is written in the Bible that the Deity had created the Earth and he sat it fixed in the center of the sky or firmament, with the entire world to revolve around it once a day. Apart from other models, if the Earth is set fixed in the sky, there is no dilemma of what is holding Earth.

This is important, for credibility and therefore relevance reasons, since other previous models of the world had to invent elephants, turtles, and even very strong deities as Atlas to hold the Earth. When asked what holds the elephants and turtles, they had to invent stronger elephants and stronger turtles to hold the previous ones along with the entire Earth, with turtles and elephants all the way down.

It seems that there was a duality of thoughts forming during the dark age, because at home in private, or among your most trustworthy friends, you could talk about a spherical model of Earth, about a heliocentric model, and about art, politics, and philosophy. While in public or at the church, you had to talk about a flat Earth and a geocentric universe. It is the same currently, if you live under a radical regime, because you have to pay attention constantly to what you say in public, while you have more freedom at home. There is always dualism in a consensual world.

Yet if you publicly chose to believe in a heliocentric model during Christianity, which is the model with the Earth revolving around the Sun, you did not choose only a simple model of the universe, but you chose an entire ideology, which was the pagan or Arian spiritual ideology of the previous age, the age before Christianity, which was dangerous.

At the time of Copernicus, people were fed up with authorities that were taking advantage of them, teaching them obedience and austerity, while authorities were doing just the opposite, living in opulence and breaking legal and moral laws continuously, without repercussions. Consequently, many people were threatening to return to the old times, which were the old times before their current religion and authorities, the old Age of the Ram or Aries, the Antiquity, Atlantida, or Atlantis. This is how the entire Renaissance movement took advantage of this exact popular tendency among the masses, it was successful, yet nothing actually changed, because the new rulers instated themselves at the top of the world to exploit and oppress the masses as it happened before.

It is said that if you do not learn from history, you are doomed to repeat it. Humanity repeated everything, throughout countless of revolutions and civil wars in each nation, when the new authorities promised everything to the masses, and after they instated themselves in power, they forgot their promises, while they oppressed and exploited the masses as all those before them.

What exactly should you learn from history not to repeat? You have to follow your own natural, intrinsic intelligent human need for freedom, and this includes freedom from authorities. Because these have only one goal, to oppress and exploit you. In order to exploit you, they have to render you weak, predictable, dependable, and obedient, and this is done only through austerity and harassment, called oppression.

Notice how, during all social movements, riots, revolutions, and civil wars, new authorities intending to take over the masses promise to the masses exactly the fulfillment of their intelligent human needs. This is why the masses place them in power, in order to be able to fulfill their intelligent human needs. Because there is something in the intelligent human needs and meanings, since they allow the development of all your abilities, physical, cognitive, and spiritual, higher and lover, as those exploiting you do not want you to develop and access them. This is why after the entire social movement, after you instate authorities in power, they forget their promises and they exploit you normally, since authorities are always part of the Consensual Matrix. The Consensual Matrix places them in power through and from the Brotherhood, while all authorities only follow assignments within the Brotherhood.

The current psychology and sociology fail to define the intelligent human needs and meanings, while they never state how important it is for humans to fulfill them. The intelligent human needs present in all humans demand continuous freedom and continuous equality and prosperity in life and in the world, while these alone can put an end to all authorities, hierarchies, and social division of the world, along with the entire artificial Consensual Matrix, if these intelligent human needs are only fulfilled. Yet there is no one coming to your door to check if you took your drugs and if you had your indoctrination, or if you fulfilled your intelligent human needs instead. Because you are free to live your life exactly as you please, and not the way stereotypes, ideologies, and beliefs demand. You can live your life at any developmental level you please, even at the servitude, addiction, and animal levels.

At the time of Copernicus, a return to the old times of Antiquity was more dangerous than believing in a different religion. Because at that time, Christianity was very unforgiving and very rigid, with its Inquisition harming and oppressing the most. Believing in the heliocentric model of the world was a sign of return to the old religion, to the old beliefs. There were people at that time switching back to paganism, or they were only menacing to do so, as a protest to Christian priests who lived their lives in sin and abundance, owning large mansions, fancy stallions, and keeping beautiful mistresses, while preaching to the masses to abstain from pleasures and sins, to live in austerity, to accept simplicity, and to fast regularly, while the masses objected constantly and even threatened to revolt and abandon Christian beliefs altogether, returning to paganism.

They did so during the time of Copernicus, they did so decades later, and they did so half a century earlier, when only Francis remained capable to teach forgiveness to the masses when they revolted, convincing them to hold on to Christianity while saving in this manner an entire religion, along with an entire world order. This is why the world cherishes Francis endlessly. Because if you are good to the Consensual Matrix, you are very famous and praised in the world.

Note that it was a heretic act only to speak against Christian norms and beliefs during the time of Copernicus, not only to publish books that challenged strong Christian beliefs, even the entire religion, as Copernicus did. Why was Copernicus allowed to behave, think, and speak in a heretic manner, and what saved him from the Inquisition so many times? Powerful, wealthy people did, who chose Copernicus the way they chose Leonardo, Bruno, and Galilei, then Einstein, Hubble, and Hawking. Who were these patrons, and why did they do so? Was it out of charity, for the art and science, and for the Masses? Again, we have to understand exactly the circumstances of those times, the scenery, and the social actors involved.

Copernicus had friends in very high places among the clergy and the wealthy, who took his side and covered him even when it placed them in dangerous situations. It was this high entourage, his friends, members of the Brotherhood of that time, allowing Copernicus to travel to the most developed cities, to have access to rare, important books and manuscripts, and to meet the highly educated people of those times. Who were the Elite and the Brotherhood of those times, over half a millennia ago?

The Elite was clearly the royalty of each city and nation, leading the world in the open, unlike currently, when the Elite have to remain hidden. It seems that all royalty in the world is gone, with one exception in Great Britain, yet there are still 25 monarchies remaining in the world, ruling 45 countries out of 175, which is a significant number. It might not be true, but sometimes, it seems that today's monarchies are not the genuine Elite, but they are only part of the Upper Brotherhood. Furthermore, The CROWN is a corporation, owned by the VATICAN, which is a corporation, owned by the CITY OF LONDON, which is another corporation, which is owned by the Families of the Upper Brotherhood, who are corporations, all the way up to the Elite. It might be different in the other monarchies of the world, because the invisible kingdom intermarried the royalty of Great Britain only centuries ago, and now it owns all corporations stated above. The invisible kingdom did the same with other Royalties of the Middle East and everywhere else, fueling the current wars. The conflicts taking place in that area have the invisible kingdom as a main reason, because the invisible kingdom is very persistent in maintaining ownership of that entire area, as it makes its way towards taking over the entire world. Yet the dictators of the East are significantly more harmful, they are not an alternative to the invisible kingdom, but only more powerful than ever, always fighting for world supremacy. Yet with the invisible kingdom always worthless, they win.

The invisible kingdom migrated in mass to Europe in successive stages, exactly at the time of Copernicus, Bruno, and the entire process of the Renaissance taking place in Europe. Coincidence? When you ask the invisible kingdom, they state that they are capable enough to have made Renaissance possible, along with the modern age and the entire human civilization past and present, since the rest of the people are only animals and therefore worthless. While they use special names for the 'animals' of the world as golem and abominations. This is what the invisible kingdom claims, just study their culture, knowledge, records, and ideologies to see it yourself.

The pagans were more developed spiritually, along with all old scientists including Eratosthenes and Copernicus, who seem to interconnect with the most developed people of their times, uplifting the developmental level of science and of the entire society. We have to understand both the consensual and the real, since both form the human environment and therefore the human reality throughout all dark ages. While the consensual seems to monopolize the actual knowledge about the human reality, and it is important to understand it in all details. We tend to study physics and physicists more, since physics studies this entire objective reality, while cosmology is the branch of physics studying how this objective reality was formed, made, created, or born.

During the time of Copernicus, the aristocracy formed the mainstream Brotherhood of those times, and served the monarchy directly. They administered and managed the Masses on their behalf, while keeping a share of the profit. They served the monarchy directly, as they personally fed and dressed the monarchy among all the usual chores, in a daily extravagant ritual.

The monarchy was genetically related to the upper aristocracy, and therefore the great part of the aristocracy kept their superior status, titles, wealth, and privileges from generation to generation. If you had not inherited your aristocratic status from your parents, you could still become an aristocrat by marrying one, under very strict circumstances. While when in need of people to serve them, the monarchy gave aristocratic status to wealthy citizens for outstanding merits during wars, or in exchange for large sums of money. Distinct aristocratic families managed the work on the land and therefore assured the food production, other aristocratic families managed trading and manufactures, while other aristocratic families led armies and fought wars for the monarchy.

The aristocracy was the equivalent of the current

Brotherhood. Everything was the way it is currently, the way it has always been throughout dark ages: the Masses served the Brotherhood, and the Brotherhood served the Elite, while most aristocratic families of the past are still in the Brotherhood today. Many non-aristocratic brotherhoods, organizations, and societies of the past are part of the Brotherhood currently, while some families from among the Masses of the past are in the Brotherhood, with some in the Upper Brotherhood and even in the Elite. Because society always changes throughout the dark ages, while there is never a rule, a norm, or an order, since it disregards all past agreements in order to make the new ones, while it only obeys the natural laws of power, cruelty, and greed. While the Consensual Matrix watches closely and only sends its orders drastically, with the Elite and the Higher Brotherhood obeying unconditionally.

What is aristocracy, and who were the aristocrats? Let us study the etymology. 'Aristo' means intelligent, noble, and worthy, while 'cracy' means a method of ruling or a regime. 'Aristocracy' means a governing regime with those in power being the most intelligent, noble, and the most worthy to rule. Aristocracy had started in the Ancient Greece and Ancient Rome, or even before them, in Atlantis, when the most intelligent and the most worthy citizens of the nation were asked to lead or rule. This is highly unlikely to happen today despite of what authorities might claim, yet it seems appropriate for any civilization of the past treasuring wisdom, spirituality, reasoning, and culture the most. These very wise people ruled their countries and nations impartially, at high efficiency, and it was specific to pagans and to the Age of the Aries.

Note the correlation between the words Aries and aristocracy. 'Aristo' means intelligent, cultivated, and enlightened, which are intelligent human qualities. The Age of Aries means the Age of Enlightenment or the Age of Reasoning. Another resembling word is 'Arian.' Arian might define a separate race of people, yet you can consider everybody on Earth to have an equal status, part of a single race, the human race. Because you can find Arians not only in the Western Civilization, but also in the East, in India and in Iran, and even in Oceania and in Africa.

As you study Arians closely, you find them very capable, very creative, and very artistic. Many of the old egalitarian Brothers were Arians and they always helped the world, while you can still find their old teachings at the lodge. All Arians seem to have migrated from the same place, the basin of the Mediterranean Sea, the place of Atlantis, which could make all Arians related or part of an initial nation, Atlantis. The word Arian relates more to the word Ares than to the word aristocrat, which might come from 'Ares' itself. In this manner, Arians are a culture, a tradition, or a civilization. While the people themselves seem to share genetic traces, being of a same origin, or being part of a same highly prosperous and highly significant civilization of the distant past.

The Arians, or at least the Arians of the past from the Age of Aries, believed in enlightenment, reasoning, magic, art, and science. If you find today specific politicians showing the sign of the devil with two fingers pointing either up or down, that is the sign of Aries, symbolizing the horns of the ram. Those are the current aristocrats, since many have survived, to be even in the Brotherhood, and probably in the Elite. Yet when you study them closely, you find invisible kingdom traces in them, since the invisible kingdom is everywhere, while genetic diversity is always sought by Life.

People are not good or bad, everybody is the same, because everybody has the same human needs and meanings to fulfill. Yet undeveloped people are mostly consensual, obeying tyrants and ideologies, while these can become very demanding, forcing you to harm those around or the entire world. This happens often, since all tyrants and ideologies fight for distinction and supremacy, as they already count in tens of thousands. Yet since the entire world serves tyrants, ideologies, and jurisdictions, you cannot judge anyone according to their beliefs, as long as you are bounded yourself by similar oaths, laws, beliefs, ideologies, hierarchic brotherhoods, and jurisdictions.

Because many beliefs, hierarchies, consensual brotherhoods, ideologies, and jurisdictions have good intentions at their base, only to do just the opposite in the world, for various reasons, since this is the contorted human reality. Now study the human mind and the human society in all details, to see how all beliefs, stereotypes, strong personal convictions, and entire ideologies install themselves at the core of your cognitive system, to judge and decide everything from there. This is how you think and behave in the world, while being controlled ideologically, minutely, knowingly or unknowingly, and it happens with everybody.

This is the first servitude developmental level, when you serve ideologies, hierarchies, and entire jurisdictions and authorities. The zero level refers to addictions and entertainment. The second developmental level is the animal level and it includes all your physiological needs along with your social, security, and recovery needs. While the third developmental level is your intelligent human level, and it includes your learning and developmental needs, along with your needs for social justice, social harmony, and social prosperity. This is the difference between people, and it involves development and cognitive abilities more than genetic, national, social, or cultural differences.

The Age of Aries was determined to develop people and to interconnect them positively, equally, harmoniously, and prosperously. This leads to a natural, progressive genetic homogeneity everywhere, making all Arians related in everything that they are. This is significant, since if you have Arian genetic background, which is the case if you are from North Africa, Europe, India, and other parts of Asia, your primal intelligences still interconnect in that specific egalitarian, harmonious, developed, prosperous, successful manner even currently, forcing you through all their needs to do the same in the world, develop it as it has always been in the past towards those golden ages, and this is what you do. While addictions, servitude, and an entire consensual environment push you the other way, ruining your life and the entire world.

We also notice different developmental levels within the genetic roots of the invisible kingdom and some other people around, generating all conflicts, since the invisible kingdom has not experienced golden, harmonious, egalitarian ages in the past, and now seeks only profit, destruction, trickery, exploitation, and eradication. Furthermore, through the continuous genocide taking place since the invisible kingdom took over the world, the specific old, prosperous genetic lines of Earth are removed systematically, with the Brotherhood and the Masses taking them down themselves. Themselves against themselves. Because the invisible kingdom is not allowed to go to war, and it is not even allowed to work, under many circumstances, escaping responsibility. You do not even see the invisible kingdom actively part in this continuous genocide, but only the people of Earth seem to harm and kill the people of Earth, continuously.

When you study important general knowledge as reality, humans, meaning, genetics, ideologies, history, civilization, science, life, intelligences, cognition, and jurisdictions, you manage to understand everything consistently from all perspectives and at all levels. While if you are developed at the third intelligent human level, you understand everything in its true accuracy, which is an achievement. While at this specific point of our mental model of the human reality, we can distinguish with ease the multitude of upper and higher forces and factions of forces spreading their power, interests, and influence on Earth and humanity through the Consensual Matrix as they always do throughout dark ages here on Earth, and in most of the wider world. While we can see large chunks of humanity fighting each other, as the invisible kingdom fighting and eradicating the rest of the world including the Arians, with these doing the same currently and throughout the history, while all interference from the Consensual Matrix cannot be blamed, because the Consensual Matrix is flawless.

Who are these upper and higher forces? Just study the deities that the invisible kingdom, the Arians, and the dictatorships from Asia worship, since they are not the same. These can give you a general idea, yet everything is highly complex above and beyond, with the entire Consensual Matrix making them even more complex, as it contorts everything in every profitable manner. While as seen, with you kept underdeveloped continuously, lacking your independent reasoning and higher abilities, while lacking even the fulfillment of your natural needs and meanings, you do not stand a chance.

Did the main scientists attempting to model and define the human reality ever stood a chance? Not even them, yet with the most capable humans united on the same goal, they managed to make an influence in the world, for good or for worse. While at this specific time and place of our mental model of the human reality, we notice how some powerful people sought a return to the golden age of Aries, while in the meantime, the invisible kingdom newly instated in the Western Civilization, achieved its gradual influence on religion and aristocracy, while seeking to take over the entire world. At least did Copernicus ever stood a chance?

You know the history, yet we can go even further in the past, at the confluence between the old golden age and the new current consensual civilization, to see what happened then. What the new religion, Christianity, feared the most, when it had been gradually implemented, two thousand years ago, was a return of its people to the old way of the pagans and Arians. This is why Christianity changed radically its codes and moral values away from reasoning, art, philosophy, magic, music, and dancing, associating them to evil or to the devil, while enforcing an unconditional belief in the Deity, an unconditional thrust or faith. The devil itself looked and still looks as a goat, closely resembling the ram, Pan, or Aries, the deity of the Age of Aries. As stated, if you happen to see high politicians, wealthy investors, and members of highly placed families displaying with two fingers the sign of the devil, it is not because they are the devil themselves, or probably they are, but that sign is the sign of Aries, the ram or the goat, depicting the horns, the symbol of the past age, the Age of Aries, the symbol of enlightenment, true science, reasoning, art, and true humanity. These symbols do not exactly point to the old times, but these are important symbols of the current Brotherhood, an entire social class having its roots in the old golden ages of Earth, in the Age of Aries, and through it, in all previous ages. This is why some factions of the current Brotherhood still point to these old times of enlightenment, wisdom, art, culture, and spirituality, since some of the current legal Brotherhood still treasures knowledge and spirituality the most, including higher knowledge.

At the first consensual servitude level, you do not only serve diligently those above while you are served similarly from below throughout very tight hierarchies spanning the Brotherhood, the Masses, and the entire society, but you also plan, conspire, and associate with others alike into hidden factions while seeking to overtake the factions above for a larger profit. While you remain caught in this continuous social competition, you use all means that you can, and this is how you are ready to rely on all types of ideological control, old and new, even on the old Arianism ideology if it is the case, if they are strong and worthy enough to make you be all that you can be. While you do so even if you have to wreck the world in the process, since this is what underdevelopment does to you and to the entire world.

This is how the Brotherhood is still divided into distinct factions, orders, cartels, movements, axis of power, and entire wings of interest, plotting against themselves depending on cases, while always managing to overthrow the world order exactly at the end of each age of Earth, in a conspicuous coincidence. People against people here on Earth, with the Consensual Matrix flawless, because the Consensual Matrix defines the entire consensus itself, always making it flawless according to itself.

The invisible kingdom is not the Brotherhood, but it controls the Brotherhood tightly from within, always in a discriminatory exploitive manner. All Brothers are significantly more capable than the invisible kingdom, yet in order to free the world from the old European Aristocracy, the old egalitarian Brotherhood decided to serve the invisible kingdom towards a new world order with the invisible kingdom in control, because the invisible kingdom promised freedom, equality, and prosperity, exactly what the old egalitarian Brotherhood sought. However, the invisible kingdom uses trickery continuously, and rendered the world discriminatory, exploitive, and tyrannical once they took control, despite of all their promises. Currently, the Brothers of the West turn away from the invisible kingdom while serving the dictators of the East for a better profit, while changing the world order once again, from the invisible kingdom to the tyrants of the East. While if you want to know what comes next, study closely all dynastic medieval dark ages of China, since they last continuously throughout its history, because the dark ages never end in a consensual world, regardless of who is in control.

The invisible kingdom is divided into wings, factions, and forces of interest, as the invisible kingdom remains divided into its distinct genetic ancestry, its main genetic wings of power. While the invisible kingdom treasures its past strategies of taking over the world, it had to infiltrate the world and therefore it had to bind genetically with the entire world. Currently, while already owning the world, its main genetic wings of power fight among themselves for supremacy, keeping the fight within the invisible kingdom, while losing the world. You can identify these with ease, since the Arian wing of the invisible kingdom remains more influential, challenged by the pure, original genetic wing of the invisible kingdom, along with the East European genetic wing and the Asian wing. While through these genetic wings of power, the persisting upper and higher forces and factions of forces still manage to control humanity through the Consensual Matrix, since this happens to be the human weakness.

Who is good and who is bad in this continuously dreadful human interaction? Are the aristocracy, the invisible kingdom, the Russians, and the Chinese bad? Is the Consensual Matrix bad instead? As always stated, you cannot divide the world into the good and the bad, but only into the developed and the undeveloped, while you cannot blame the undeveloped, because with the entire world undeveloped and mostly in dark ages, you end up blaming all humans with you included, throughout most of the human history. While at the intelligent human level, you seek to understand the entire circumstance involved, while trying to remain developed, fulfilling, and harmonious in life and in the world.

Therefore, at the intelligent human level, you do not seek blame and revenge, for whatever people did centuries, millennia, and ages in the past, because this is currently irrelevant. While you do not even seek to join this everlasting human fight within your nation, Brotherhood, or entire world, since even these remain irrelevant, through the continuous harm that they make you inflict in life and in the world if you remain undeveloped yourself. Since while undeveloped, it does not matter where you join this fight, because you end up in this continuous genocide yourself, while harming life and the world more, even endlessly. Because at the intelligent human level, you seek to maintain and enhance harmony and development in the world, by developing and remaining harmonious yourself, while influencing those around through your own harmony, development, and fulfillment, including your loved ones. While as you notice, you have to know everything in order to do so, about humans, life, reality, intelligence, development, and even about Copernicus. While you have to remove all first level consensual elements form your cognition, life, and social interaction, as ideologies, jurisdictions, and hierarchies, in order to live your life only in the living real world, in the real human reality.

Yet currently people choose drugs, servitude, and tyranny instead of an entire intelligent human fulfillment, while making the current dark age last indefinitely. It is always a difference between what people do and what they were supposed to do, and this is how currently, many people are drawn to drugs, profit, and exploitation, harming the world even badly. While this has nothing in common with the old ages of earth, with the devil if this even exists, or with entire past Brotherhoods more or less egalitarian, since people are interested mainly in drugs, social power, and tyranny, but not in old mythological characters more or less egalitarian or diabolic. Yet if believing in all these demons helps them have even more drugs and social power, then they can display all diabolical signs in the media whenever demanded.

Not only humans decide everything in this world, but their souls also do, and these have their own cultures, brotherhoods, ideologies, and strong higher convictions, with the Consensual Matrix reaching them more, since the souls bring the Consensual Matrix here with them every time they come down. While many higher selves reason intelligently, interconnect harmoniously, and have higher abilities, influencing the world directly and many times in a good manner. While other souls come to this world for drugs and tyranny even combined, in order to have a blast while caring less of this entire world.

If groups of smart, cultivated people already exist, seeking to help humanity, then this is very good. While if there are groups of people that use knowledge and enlightenment in order to profit and oppress others, then this is dreadful. It is the same with the Masses, since masses will always exploit masses if they only have the chance and if they live life undeveloped. While masses will also exploit brotherhoods and elites once they have the chance, since it makes no difference when it comes to consensual profit while living life on lower developmental levels.

Some people show the sign of Aries inverted. These people from the higher layers of society can also descend from the old aristocracy, with some of them descending directly from kings, queens, emperors, and pharaohs, while others only claim royal ancestry.

It seems that some of these are the people that turned the Western Civilization slowly away from the dark ages, from the rigid, unconditional religious beliefs, towards enlightenment and reasoning. They have done so in a matter of centuries, which is an achievement. This change in the mode of society after Renaissance and up to the present day might have happened not as a return to the Age of Aries or the Age of Reasoning, but as a full emerge in a brand-new age, the Age of Aquarius.

In general, inverted ideological signs and spiritual signs symbolize the end of an entire regime, age, dynasty, ideology, or social class. While when you display the sign inverted it signifies your agreement to end that old age, regime, ideology, or political party, and start the new one. Because all flips in the paradigm are very well organized, and it always matters to state your own side in the entire social organized fight, or you are hurt unnecessarily.

As already stated, in some nations of the last age, the genuine aristocrats, which were the wisest, the most enlightened citizens, were asked to lead the entire nation, being poor or rich. This governing method had brought prosperity in the nations of the past age, since it eliminated corruption and discrimination, among other issues. However, in time, these genuine ancient aristocrats learned how to create wealth, power, and influence throughout their term, they learned how to remain longer in power and how to keep this power in the family, until they formed genuine dynasties and entire dark ages. Coincidentally, this was exactly what led to the destruction of their outstanding nations, including Atlantis. Social islands within social islands means discrimination, marking the end of the harmonious egalitarian golden human age. The aristocracy and the monarchy were reborn to lead openly the Western Civilization, or this is what the old Brotherhood claimed, and they are still here. While in the East, dynasties ruled unconditionally continuously, one dark age after another.

Even the smartest and the most well intended people of Earth cannot instate and maintain an entire civilization harmonious if they are not developed themselves, dropping it entirely in tyranny, exploitation, and dark ages. You cannot govern people through laws, beliefs, authorities, hierarchies, ideologies and jurisdictions, because these are of the first consensual level, while humans are third level intelligent living human beings, remaining incompatible with laws, hierarchies, authorities, and the entire consensual.

Yet laws, beliefs, ideologies, and jurisdictions are always instated in the world even with good intentions, through ignorance and indoctrination, since only through ignorance and indoctrination, you are capable to control those below, and you can do so only for as long as they remain underdeveloped. This is why you always see sickness, misery, ignorance, loss, and suffering in the world, since these enforce and maintain underdevelopment, maintaining compatibility in this manner with the first consensual, servitude, ideological, juridical level. Then once you have underdevelopment in the world, your laws and ideologies compromise people's cognition, and you take it from there, since they will always be willing to follow you as long as you assure their social competition, addictions, discrimination, fights, revenge, and exploitation.

Why was part of the aristocracy of the time of Copernicus interested in implementing a heliocentric model of the world? It seems that they were not interested in the heliocentric model of the world in particular, but they were reinstating all culture, knowledge, norms, values and lifestyle from before Christianity. The age of Aries seemed very appealing to aristocrats and Masses alike, tired now of the very low lifestyle that religion was maintaining. Promises of return to the lifestyle and civilization of the previous age were sufficient to assure the support of the lower social classes to any opportunistic powerful family or faction of families to take over the world.

Those were the times of the Renaissance, and they were just starting. It was the return to art, science, and reasoning, allowing many aristocratic families to become very wealthy, doing business in banking, trading, war, and manufacture. This was enough to unbalance the social structure of that time, since wealth and power alone shape society in every civilization that is not based on reasoning and harmony but on crude social hierarchy and competition. These new wealthy families and factions of families had worked hard to acquire their wealth and power, and were challenging the upper social layers, all the way up to the monarchy. Entire factions of power acquired significantly more wealth even than those above them in society, since new ideologies and new technology always offer all means of wealth and social advancement.

For example, the De Medici, which were the patrons and protectors of Leonardo da Vinci at the time of Copernicus, were bankers, traders, and miners, similar to other highly prosperous families of those times, challenging with their increasing wealth every family from the social layers above. Was Renaissance alone capable enough to change an entire paradigm and create a new world order? Yes, definitely, since it made he De Medici family the wealthiest family of the Western Civilization, comparable to the Rothschild family of the following centuries. The Rothschild family had also acquired wealth and power while ascending socially in a similar manner, by feeding on the discrepancies and disagreements among the monarchies of the Western Civilization, by feeding on the Industrial Revolution itself, and then by feeding on the two world wars while overtaking entire factions of remaining powerful aristocratic families, in order to ascend socially all the way to the Elite and take over the entire world, taking it away from the old European aristocracy itself.

Yet as stated, the old egalitarian Brotherhood chose to serve the newly emerging invisible kingdom, in order to save the world from the old European aristocracy, because the old European aristocracy were keeping the West and the entire world in a continuous dark age, and wanted a return to the old golden ages of Earth. This is why they helped Copernicus in science, and this is why they helped all scientists, artists, teachers, and scholars alike, because it seemed as a plausible return to the old golden human ages, yet it never happened, since the invisible kingdom never kept their promises.

The Rothschilds were royalty, served minutely by their

entire invisible kingdom, and this is how they took over the world. The Rockefeller family had always served closely the Rothschild family, and was pulled up the social ladder along with it, along with many other faithful families, which are currently in charge of the world. Our past centuries had been a turbulent time for the upper social classes, significantly affecting entire factions of families, some ascending and others descending, throughout a true powerful inner war and a true vivid existence. While the Masses remain currently aware only of isolated names in history as Louis, Napoleon, Lincoln, Hitler, or Churchill, stating only these while discussing history.

The three families stated above, De Medici, Rothschild, and Rockefeller, are not isolated cases or outstanding examples of social achievement, since society is in a continuous change and in an unending development, while what happened to some families along the way will happen to all, eventually. This is how members of the De Medici family became popes and queens, while currently, you can still find their genetic traces as high up as the Elite, mixed with genes from the other two families stated above along with many others remaining hidden, as everything relates with Copernicus and the rest of capable people of Earth making all intelligent human achievements possible, because the upper social classes are worthless, and cannot invent anything on their own. Yet all tyrants and megalomaniacs are worthless, with the entire invisible kingdom included.

The entire world praises our very capable artists and scientists not because the entire world finds them worthy, but because the higher societies and the Consensual Matrix find them worthy, to them, since these are in the Brotherhood, serving them well. While the world is full of capable people and souls, past and present, always willing to make the world a better place, yet they are shadowbanned and marginalized and you never hear of them, because they are not in the Brotherhood, and they cannot be controlled. Copernicus, Leonardo, Verne, and Descartes are enough, no more, because you can always control these since they are in the Brotherhood, and therefore no more.

While as you study the Age of Aries closely along with all golden human ages before, you notice how everybody was a successful scholar, teacher, thinker, artist, or scientist, since only in this manner, you can maintain your intelligent human developmental level, while making the entire world intelligent and harmonious. Yet by decaying the world deliberately in order to control and exploit it, you render everybody undeveloped throughout all possible dark ages, while ruining an entire human world.

Copernicus had continued his research of the heliocentric model of the world, published his books, taught that the Earth revolved around the Sun, he never stopped at the threats of the Church, and one day, whenever the other wing of the Brotherhood became stronger, they burned Copernicus along with all his books, and they banned all his studies and ideas. Yet it was too late, because his scholars continued his model of the world, and never let it go. It was the same in art, music, architecture, and even in business and trading, since the lower classes returned more to the lifestyle, culture, and beliefs before Christianity, yet they never went back to the Age of Aries, which was astronomically impossible, but they developed their knowledge of the universe, along with their culture and civic structure.

The model of the universe implemented or re-implemented by Copernicus was taken and carried forward by Giordano Bruno, Nicholas Kepler, Galileo Galilei, and later on, by Isaak Newton.

Giordano Bruno was born when Copernicus died, in 1548, in Naples. Bruno was a genius, being able to memorize entire books by using an ingenious cognitive method that he had perfected himself. This might seem irrelevant, yet if you happen to have this ability of accurate memorization currently, you can easily earn your medical and law degrees whoever you are, since the current education is based only on memorization, not on art and intelligent reasoning.

However, it was different with Giordano Bruno, because

during those times, books were very rare and very valuable, through the information that they carried. The printing press was not invented yet, or this is what science states, making all books very rare, and it was always helpful if you could memorize entire books, since you kept them in your mind.

Later on, the printing press made information so well available to people, that it started the entire Age of Industrialization, leading the world to be as you know it today. While currently, the Internet makes all exoteric knowledge available to everybody, and this leads this entire Age of Social Interconnectivity, which is very similar to the socialism ideology, but at lower developmental levels, since ideologies are of the first consensual ideological level.

If humanity develops to its third intelligent level, this Age of Socialization leads to the age of the intelligent human family, or the age of intelligent human living, which is the overall human family. Because when you study history closely, this is what humanity has always sought, the return to the overall intelligent human family, the intelligent human communal life, with religion and the rest of ideologies and jurisdictions always standing in the way, demanding and enforcing these small human families that you currently have. This is the difference between the human natural reality and the human consensual reality, the ability to live your life in overall intelligent human families, groups, and brotherhoods, and not only in these very small families or in no family at all, as it is the case towards the bottom social layers of the current society. Yet you cannot have the genuine overall human family and the Consensual Matrix simultaneously on Earth, since nobody listens to consensual authorities anymore within overall intelligent human families, because no one has to, since your overall human family and the entire egalitarian Brotherhood tend to your needs just as you tend to theirs, and this is the third level intelligent human reality, what humans should always have.

This is what people wanted continuously throughout the centuries and millennia, this intelligent human living, and they knew exactly what to ask for, the overall human family, the overall human commune, or the overall communal living. As they did in Paris decades later, when they demanded to have the Commune of Paris instated or reinstated, and they failed, unfortunately.

Because you always fail developing the entire world with you in it when you are underdeveloped yourself and when the others remain underdeveloped alongside you. People might have kept wondering why they cannot reinstated the old golden human ages, while blaming tyrants and entire ideologies, yet people tend to ignore that you cannot have drugs, ignorance, and golden ages combined, for lack of compatibility.

More precisely, people wanted to have ideologies, tyranny, drugs, harmony, prosperity, and communal living combined, which is impossible, because you must avoid drugs first while developing continuously in order to achieve golden human ages. Furthermore, people demanded harmonious intelligent communal life from their own tyrants, while their own tyrants decided that it is better to keep tyranny, social disconnection, ignorance, and austerity instated, and the people obeyed.

People always remain unsuccessful when they wait for their authorities to give them privileges, which they use in place of human rights, while all authorities are part of the Consensual Matrix, and they will never consider human rights, because human rights are part of the intelligent human reality, while privileges are part of the consensual human reality, and there is a significant difference between the two. This is why you never learn these from the current sources of knowledge, otherwise you actually develop, and you remove the Consensual Matrix from Earth.

As it happened before, when they kicked out the Consensual Matrix at least by turning their back to authorities and therefore to the Consensual Matrix. This happened continuously throughout many past ages of Earth, yet people used higher abilities then, and were always successful. While currently, all higher abilities are systematically removed from everybody, along with entire people, genuine egalitarian brotherhoods, and entire very capable genetic lines, only to keep the Consensual Matrix on Earth, and it is still here.

Yet how can reality be different, sometimes real, natural, human, and alive, and sometimes only fiat, consensual, and tyrannical? Reality is reality, and it is only one, since existence defines it uniquely. However, you can always agree with others on anything that you want, imagine, need, and desire, since this is how the unreal becomes real, the lies become true, the bad becomes good, and the yellow becomes blue. This is the consensual, making everything possible, but only consensually, in a fiat manner, not actually there.

You do not make anything real according to your agreement, but you only consider it the case together, it is still not the case, yet you act as though it is the case, in a rather ignorant manner that ruins the world. Let us now consider together that Kim is the most wonderful and most capable man on Earth, which is not the case in the real world, yet with the entire North Korea assuming so together, Kim becomes and remains the most wonderful and the most capable man at least in North Korea, which still matters, while ruining the lives of all North Koreans.

Yet isn't it better to live life in agreement than in disagreement? Don't you have chaos and destruction through disagreements? Isn't this the devil? Yes, but you have to see further, since you always have to agree on everything that is not already the case, on everything that is not already real, otherwise you never had to agree on it, because it was already the case.

Let us agree now that the sky is yellow. This is how we form the agreement and the belief, that the sky is yellow, which is not actually the case, because the sky is blue. However, every time we meet, we state that the sky is yellow, we believe it strongly, and for us the believers, the sky is yellow, yet only for us. We can form an entire ideology around it, and this is our life from now on, under a consensual yellow sky. This is how reality is always changed consensually, while it is supposed to be accurate and unique. Furthermore, we can form an entire jurisdiction around us, defining clearly in our statute that the sky is yellow and it becomes so by law, with the necessary punishment instated whenever our people break this law in any manner, since this is how jurisdictions are created consensually, now covering the world.

While you never have to agree and make laws that the sky is blue, since it is always blue, while this is the consensual compared to the real. The consensual is never real, yet since the consensual is never real, it can always be contorted in every manner, just ask the lawyers to tell you how it is done, according to all tyrants controlling them. Now this is life and this is the world, consensual, affecting humans continuously and the entire wider world, while defining the human reality, crooked and contorted.

This is how you end up with various levels of reality and of existence, while these are supposed to be unique, always accurate. This is how the first level consensual reality is only agreed upon but it is never real, only a code, only a play, only an act. You find these words and concepts throughout all legal procedures, but you lose your real house, your real wife, and your real cat in this manner, considered only an act of law, a play, or a code, but nothing real. This is the first level consensual human reality, with you as a corporation at the first consensual level. Since this is how you are accepted in court and throughout jurisdictions, at the first consensual level, as a corporation, but never as a real intelligent living human being. You are even screamed at and removed from court if you persist to state that you are a living human being, because they cannot have you in court if you are not a corporation, maintaining compatibility. The court itself symbolizes a consensual place apart from the real world. The consensual is always of the first agreed level, while life and the actual real world start with the second level, while within life and the real world, humans are at the third intelligent level by nature.

During the times of Giordano Bruno, books were still not printed but they were handwritten individually, and they were still very expensive. You had access to books not only if you purchased them, since books were a privilege, more expensive than shoes and horses, yet if you had access to books at least occasionally, as during social events, when book owners allowed you to admire their book collection, if you could memorize them then, by reading them very fast, and you could recall them later on as best as you could, to enjoy your own book collection.

Giordano Bruno entered the Dominican Order and became renowned for his art of memorization. He even travelled to Rome, happy to demonstrate his outstanding talents to Pope Pius V, while the Pope was more concerned with Bruno using sorcery to memorize everything. Giordano Bruno wrote a book later on, dedicated it to the Pope, and gained in this manner his patronage, which was a high achievement.

As expected, Giordano Bruno developed a taste for rare, forbidden books. He was interested in old knowledge, anything from Arab astronomy to Egyptian mythology. One advantage of memorizing every book that you read is that you have all information that you ever need at your fingertip. Giordano Bruno could also write his books in his head, and he dictated them to his assistants later on.

Giordano Bruno had outstanding views of the universe. He followed in part the Copernican heliocentric model of the world, while the Church had already found it inaccurate and already banned it. Yet Bruno got away, because he was capable to escape jurisdictions whenever necessary, since during those times, people lived in cities and not exactly in states. This is how jurisdictions span only cities but not entire nations, and you could simply move from one city to another to escape jurisdiction.

Giordano Bruno improved the Copernican model of the world considerably, and made it plausible even by current standards. While the Copernican heliocentric model stated that the Sun was the center of the universe, Giordano Bruno considered that the Sun was nothing but another star, while the universe is infinite with stars, going as far as you can imagine, since the Deity is infinite, his powers are infinite, so it is more likely that his creation is infinite, or this is what Giordano Bruno stated. Furthermore, Giordano Bruno stated that there could be planets around all stars of the infinite universe, and therefore humans are not the only intelligent living beings in the universe. These were very dangerous affirmations for those times, since by claiming that there are other civilizations out there and we are not alone in the world, it causes you problems even today.

We can wonder now why the Inquisition burned Copernicus for less heretic thinking, while allowing Bruno to live. Yet the Inquisition did not burn Copernicus for his heliocentric model of the world, even though the entire world assumes so currently, because they could not. There is nowhere in the Bible stating that the Earth does not revolve around the Sun. In the Bible, it is stated only that the Deity had created the Earth, then He created all plants and animals, then He created the men to rule above all animals everywhere on the surface of Earth, and then He sat everything fixed in this manner, with nothing to be changed. This means that nobody is allowed to interfere with His decision and with His creation, but he did not exactly state that the Earth is physically fixed in space. While as you notice, all higher beings interfere with Earth through the Consensual Matrix against the wish of the Creator.

Giordano Bruno knew exactly how to behave in order to remain safe. He knew how to find and read forbidden books, how to teach his knowledge, and how to write his own books. Bruno tried not to interfere with the Bible and with the Christian beliefs in any way. He certainly knew the Bible by heart and was smart enough to know how to interpret it more than the high clergy, while their interpretations during those times must have insulted his intelligence. Bruno had always remained within Christian beliefs, making the inquisition work hard to find anything to condemn him. While when they found something, Bruno moved away to other cities within Italy, and then he moved to England, France, and Germany. He managed to dedicate his books to kings and to high dignitaries, gaining their patronage and support. He taught in universities when they accepted him, and he wrote many books.

Giordano Bruno was a genius with connections in very high places, yet he was not too successful in his teaching career, and he still could not find a teaching position. He was about forty years of age when he obtained an invitation to move to Padua, to tutor privately a very high dignitary, with the promise of obtaining the Chair of Mathematics at the University of Padua, along with the assurance that the Inquisition forgot about him. Yet the university rejected him but hired Galileo Galilei instead, while the Inquisition did not forget about him.

Two months later, the Inquisition arrested him, imprisoned him, convicted him, and then burned him on the stake, in 1600. On what charges? Giordano Bruno had made three affirmations either in writing or verbally. He stated that Christ was a stronger magician, that Virgin Mary was not virgin, and that the Holly Ghost is not in each one of us, but it is the spirit of the entire Earth.

Study these statements closely, to find them related to old Earth spirituality considering life and the living Earth, nature, and humanity more than consensual agreements, as you always find a continuous tension between the natural intelligent human reality and the consensual human reality. You can never gain your freedom from the Consensual Matrix, not only because of the human greed always standing in your way, but because you lack the higher abilities that are already part of the natural intelligent human reality. Bruno had his higher cognitive abilities helping him promote the genuine, natural human reality, yet without much support from everybody else, he gave up his fight for freedom and intelligent human living, preferring death instead. Because he could have escaped jurisdictions indefinitely, yet he preferred to die in his native land.

Because you cannot have your genuine natural human life, environment, fulfillment, and reality, unless you have the entirety of your human cognition, including your higher abilities, which are currently removed systematically from everybody, through drugs, chemical additives, indoctrination, divertissement, addictions, and underdevelopment.

Ask anyone who our martyrs of science are, and they will mention Copernicus, Bruno and Galileo, burned at the stake for their determined scientific convictions that Earth is not flat and it revolves around the Sun. Yet both Copernicus and Bruno had been convicted and burned for their religious statements that went against religious beliefs unrelated to science, while Galileo was convicted for heresy and he was not actually burned, but he was only confined to house arrest for life, since he had abjured all his affirmations against the Church. More precisely, he denied, cursed, and rejected all his heretic words, he stated it clearly, and he did not die but he was only imprisoned, for life. Not in a dungeon, but in a mansion, at the residence of one of his friends, an Archbishop, and then later on, in his own home.

Galileo was named 'the father of science,' 'the father of physics,' and 'the father of astronomy.' What contribution did Galileo bring to science? He had made many astronomical discoveries, as the moons of Jupiter and the phases of Venus, and he had supported the heliocentric model of Copernicus. Why not the model of Giordano Bruno, which still seems suspiciously accurate for that time? The answer is unknown, and nobody ever asks. Why was Galileo really the father of science, of physics, and of astronomy? Galileo was the first to state that there are no fixed points in the universe, that everything is relative, all depending on where you place the origin of your system of reference. This was the idea leading to the discovery of the natural law relativity.

Furthermore, study all these historical figures, to find them strongly involved in the Brotherhood of their time, with those with a higher status in the Brotherhood having priority over those with a lower status back then, maintaining this precise order even currently, through their current recognition, long after they died. Because the Brotherhood can certainly remember and cherish you if you had served it well, and mostly if you had served it better than others. This is the Brotherhood attitude determining many to serve at all costs, regardless of the harm that they inflict in the world. This can lead to a strong social competition favoring the Brotherhood always from within, and it makes the Consensual Matrix very strong and very capable over Earth.

Notice the obvious division within the Brotherhood of those times, between those who sought a return to the intelligent human civilization of the old age, the Age of Aries, and those who wanted to maintain the world as it was. This was not a simple choice based on culture, reasoning, or world prosperity, but it was simply an internal social battle for supremacy within the Brotherhood and in the entire world. Because with each world order, entire factions of the Brotherhood overtake the rest while replacing even the Elite at times, with the weaker faction awaiting in the shadows of the Brotherhood for any significant advantage to help it recover to replace those above and to rule the world once again. While trashing the entire world in the process, adding to the dreadful human condition.

Currently, technology can offer a significant advantage to overtake others within the Brotherhood, the way industry, computers, Internet, commerce over the Internet, and social interconnectivity over the Internet allow many Families and factions of Families to climb in the Upper Brotherhood, replacing the weaker factions. While back then, the Families of the Brotherhood had to rely on the Masses of those times and on their support to achieve world supremacy, promising even a return to the better old way of life, as it is still the case today. This is how Copernicus, Bruno, Leonardo, de Medici, and Galileo helped, and this allowed the entire Renaissance movement to succeed at least partially, while bringing some Black Nobility within the Upper Brotherhood, replacing some less successful aristocratic nobility in the process.

Yet the spirit of the Masses and of all souls went on, they studied science and performed art, since it was relatively more allowed after the Renaissance, as promised. They instated the

Industrial Revolution first, and this gave the Western Civilization priority in the world. The invisible kingdom took advantage, then the World Wars followed, also on their advantage. The invisible kingdom changed the entire world order of those times, replacing the aristocracy and royalty with their own kind, and now this is the human reality, actual and consensual, with you in it. While we could have a wonderful intelligent human world instead, if you lived your life in life and in the real world.

Therefore, it is not exactly your extraordinary work, talent, reasoning, and high knowledge helping you succeed in the world of art, business, politics, and science, but you must be part of an entire movement within the Brotherhood of all times, and within the winning world order, in order to gain world recognition. Otherwise, whatever you manage to invent, discover, write, and paint is stolen by the cruelest and by the most zealous of the Brotherhood and the invisible kingdom, to be used on behalf of the entire Consensual Matrix, or it remains forgotten, because the Masses and the Brotherhood are trained well never to recognize you and your outstanding achievements if you had not served them well.

Currently, books, scientific research, and genuine art remain obscure and even censored, only to allow a large amount of trash and entertainment to gain popularity as art, because you had not served the Brotherhood enough, at your best. This is yet another source of all evils, being related to blind, unconditional servitude that is even called virtue, and it engages everybody, in a tremendous social competition, fueling the Consensual Matrix, since it determines people to succeed at all costs, even by harming the world, as we see it happening throughout all wars, depressions, pandemics, and the rest to the major consensual disasters.

Therefore, it does not exactly matter how capable you are as an individual human being, but it matters more the entire consensual agreement within the hierarchies of the Brotherhood, since the Brotherhood always overtakes you. Furthermore, higher beings are always involved, and their higher contribution certainly matters, along with their own consensual harmony or collaboration within higher and lower higher organizations. We notice a division of the entire wider world into the living and the consensual, constraining all living beings, higher and lower, as these wings of power rule Earth from one age to another, determining an exact order, culture, spirituality, or religion.

If you want to take your own stand in all these, only not to miss your chance for power and recognition, know that this is not exactly a simple political event or sporting event, but you have to take your place and participate yourself in the extraordinary milking and churning taking place throughout the entire wider world, alongside any two major wings of power, where you simply milk and churn yourself, your loved ones, your cartel, your faction, and the entire world, and where everybody does the same, since everybody is in it. This divides our model of the human reality into the consensual human reality and the actual human reality, and it is important to know it.

It was the time called the Renaissance, the renaissance of the civilization, culture, tradition, and spirituality of the old age, before Christianity. Could it have been a natural spiritual evolution of our civilization happening at that exact time, the way it happens today? Because people make new inventions in science everywhere and every time, and they create works of art, coincidentally similar to the art of the Ancient Greek and Ancient Rome. If this is true, then the Renaissance, which was that entire movement meant to end the dark age of Europe, would have been called 'The Naissance', or 'The Birth', but not 'The Renaissance', or 'The Rebirth.' Besides, a naissance, or a birth of an enlightened age in the middle of the Age of Pisces is impossible astronomically, and therefore the dark age was doomed to continue until the end of the Age of Pisces.

China had a great wall around itself, and it was tightly controlled by powerful dynasties, which were served unconditionally by a rigid Brotherhood, and therefore nothing got in and nothing got out. The Chinese model of the universe was very old and very simple, with a flat Earth and the entire universe revolving around it daily, until the seventeenth century, and it was the same in Japan.

Renaissance was good news for the Western Civilization. However, the old Age of Aries never came back, since it had never been intentioned to come back, but it only facilitated a new world order, making some families extremely rich and powerful at the expense of others within the Brotherhood. Christianity kept on ruling, and they even burned witches and pagans, they burned members of other religions who refused to leave, and they burned heretic scientists, including Copernicus and Bruno.

The consensual Brotherhood is possible only throughout entire world orders defining minutely its hierarchies of power. These world orders change drastically from one age to another, with the new world orders replacing the old world orders. The new world order of the time of the Renaissance needed to control the Masses, just as the old world order did before them. Therefore, they kept Christianity to control the people after the Renaissance, while people's dreadful conditions continued throughout the new world order, now called the Age of Enlightenment.

Yet there are other reasons regarding the presence of religions in our society and civilization, reasons escaping our comprehension right now, so let us study them right away.

6 OLD RELIGIONS, NEW RELIGIONS, AND THE ACTUAL DIVINE

When you study the current science, you understand only this consensual world, but not the actual wider world, everything above, because the current science refuses to leave the perspective of this world throughout its research. Furthermore, religion and spirituality should also offer a model of the wider world, including all planes of above existence, since religion and spirituality should put humans in contact with the spiritual and the supernatural, with the unseen and the unknown. The largest religion in the world was Christianity, and we are going to study Christianity as a model. I am not discriminating among religions, since everybody has the right to believe, venerate, learn, and ascend spiritually as they please.

Christianity is not the genuine religion of the Age of Pisces, since the genuine religion of the Age of Pisces is Coptic, currently either deceased or covert. If you ever see unknown symbols in Christianity, as the shape of the Pope's hat resembling a fish, making him represent a fish, among other symbols and representations, if you ever study and notice some people at the very top of Christianity behaving differently, seeming to follow a different religion altogether, it is important to know why.

Three thousand years ago, when vice, discrimination, and corruption overtook the nations of the outstanding Age of Aries one after another, the people became frustrated that science, art, and religion were kept inaccessible to them, allowed only to the wealthy and the powerful, within distinct brotherhoods that tended to each social domain, unwilling to share with the rest of the world, and therefore affecting drastically the entire society and causing it to fall. This had remained the case even hundreds or thousands of years later, when science and religion were not only hidden, but were kept in a different language altogether, the old languages of Ancient Greek and Ancient Rome, the ancient age of Earth. They had remained in this manner for centuries to come, mostly in Latin and Ancient Greek, while regular people did not speak Latin and Ancient Greek.

Religion was also banned to regular people, and it was kept only in old languages, until only six hundred years ago. Furthermore, the masses could not even read and write in their own language, so everything was out of their reach. The Masses had regained their rights slowly, throughout the centuries, and only during the times of King James, they had access to the Bible. It was and it still is a specifically modified version of the Bible, the King James Bible, made only for the Masses, the exoteric Bible still in use, yet it is allowed only to some Christians, not to everybody else.

Why having discrimination of knowledge, and more importantly, why is it still currently the case? It is part of the discrimination between the social classes, the Masses, the Brotherhood, and the Elite, and it is always present in society, as long as you have society divided into social classes, consensually. Because in the real human world, you are always equal and always harmonious.

Who would ever change the Scriptures, the very words of the Deity? Only those who have specific interests, since this is how you contort the human reality. The Bible had been modified before, during the Emperor Justinian, at around 400

A.D. We could not know exactly what the Bibles taught before, but rumors say that all references of reincarnation were erased right then from the Bible, among other divine teachings. Jesus became the Son of the Deity then, not the reincarnation or avatar of the Deity on Earth.

Why the change? To erase the existence of higher selves and higher worlds, to hide the act of normal incarnation on Earth, to shut down all doors of knowledge and perception leading to an entire higher existence beyond this world, and to hide all other realities from people, from the Masses and the Brotherhood alike, and probably from the Elite. Because by knowing everything, everybody would actually know the human meaning in life and in the world, and would act accordingly at the third intelligent human level, making all tyrants impossible. Yet since this is not what the tyrants want, and since people always obey them, you have one dark age after another, instead of golden human ages at the third intelligent human level and above. Knowledge allows people to ascend socially in a consensual world, with those already in power always determined to fight back by weakening, disabling, dumbing down, and even eradicating those below, only to keep their supreme social place endlessly, since this is how all tyrants are possible in a consensual world.

This still happens currently, while causing the continuous ignorance, weakness, and eradication of the Masses and the Brotherhood alike, for all these tyrants to exist. Yet as you study history closely, you notice how all tyrants never last indefinitely, but they are always killed at the end of their reign, in order to make room for the new crueler tyrants. While some tyrants are so cruel, that they are ready to destroy an entire world in order to maintain themselves in power, under all circumstances, as it is the case with Xi, Putin, and the rest.

Yet this is the case as seen from within the human society, because from the perspective of the Consensual Matrix higher above, knowledge alone can determine the humans and their souls to abandon the Consensual Matrix, or not to agree anymore to be exploited within the Consensual Matrix. This affects the tyrants of the Consensual Matrix, and it is inadmissible for them.

Currently, the entire model of the wider world is hidden and forbidden by the mainstream science, spirituality, and religion, a model including not only this world, but all worlds and all realities above, the entire wider world, which are still part of the human reality, since the human souls live there. From among all these, only the knowledge of a few other planes of existence is allowed by the major religions, as the Heaven, Hell, Paradise, Infernos, and Purgatory, while the current science hides all the worlds and realities above, while hiding all souls and all higher beings altogether, including our Creator, the Deity, and Life herself.

Why shutting down all doors of knowledge and perception leading to higher reality, higher life, and higher existence? It is as asking why kidnappers, aggressors, and harassers keep their victims confined continuously, why guards keep prisoners strictly confined in locked cells, why there are strong fences around all schools, why royal guards keep kings and queens safe while keeping unwanted people from harming them, why suppressed minorities are kept in ghettos under strict rules, and why endangered plants and animals are kept in national parks and in natural reservations.

People give many answers to why humanity is kept confined and in ignorance of the wider world, while some are right and some are wrong, yet only by understanding clearly the wider world we can find the right answer, as it is part of the actual wider human reality. Yet as seen so far, humanity is highly more profitable at its younger stage of development, as it is kept protected from the more developed civilizations, worlds, and realities, as it grows and as it continues growing, and this is why the Consensual Matrix keeps it confined, for a very long time. Whenever the Consensual Matrix decides that humanity is ready to be part of the wider world, it becomes officially part of the wider world, yet not until then, in order to be exploited some more.

Note that everybody is seeking answers in the world,

regardless of their social status, yet members of the Brotherhood are privileged to have higher answers by default, according to their degree, influence, and status. These higher answers are very diverse, sufficient to satisfy one's need for higher knowledge. However, how much weight can they put on these answers, when these answers change constantly and systematically, only giving them the impression that the more they ascend in degrees, the closer they get to the truth? You always have to learn and to understand yourself and the world, and not to wait for answers from others, since those answers might be only consensual, not accurate. While if all the answers that you seek change with each degree that you have in the consensual hierarchic Brotherhood, it is certainly consensual knowledge but not accurate, because all accurate knowledge is unique and cannot change.

Who controls the amount of knowledge that people have in one century or another? Another question is why there are two men sitting on the same horse. It has been in this manner for millennia, and it seems uncommon, inconvenient, inefficient, and a terrible burden for the horse. More precisely, why are there two separate ruling structures meant to control and manage the Masses simultaneously throughout the dark ages, the church and the Brotherhood. Most of the higher clergy were members of the very wealthy families of that time, as De Medici, who were popes, aristocrats, bankers, and sometimes royalty. Therefore, why having two horsemen on the same horse, two ruling systems in the same society, mostly when their members are from the same families? The only answer that we can provide right away is that these two separate governing systems, the church and the aristocracy, were not both serving the same master, namely the monarchy, but someone or something else. Furthermore, nobody has ever harmed Christianity, even during the harshest times, while aristocrats always hurt aristocrats and factions of power always harm factions of power throughout the unending social competition. Who exactly is religion serving? The Deity, while the aristocracy served the monarchy, making for the two horsemen on the same horse, while adding to the dreadful human condition.

Why having two governing systems? Aristocracy was clearly serving the monarchy or the Elite, and it still does so today. Could the church serve the Deity, as it truly claims? Then the Deity is real. We cannot simply guess the answer of this important question, since we need to study the church and religion in all details in order to learn the truth and give a definite answer. What is religion? How did it come to be? Is it true, or merely a myth, an invention meant to control the people, with the people already having a system of control in place?

There are tens of thousands of religions in the world, so let us study them diligently. It is believed that most religions are as old as the universe. Religions come close to explaining and creating a model of the universe, as close as science does. It seems that if we want to create a true model of the universe, we need to understand not only science, but religion. More precisely, we need to understand all religions in place at the time of all mainstream models of the world that we study.

What is religion? A pertinent, generally accepted definition of religion states that religion is a set of variously organized beliefs about the relationship between natural and supernatural aspects of reality and the role that humans have in this relationship. While all sets of beliefs form ideologies, and if these beliefs are religious, they are religious ideologies. Religions are ideologies, very similar to social ideologies as capitalism and communism, to political ideologies as communism and the red shirts, to scientific ideologies as the current science, and to national ideologies as the nazi itself. All sets of beliefs form ideologies while all sets of laws form jurisdictions, yet jurisdictions are only legal juridical ideologies.

Studying all religions, we find two main traces, two main religious tendencies. The first type of religions, more common in the past, helped people understand, get involved in, and master their relationship with the supernatural aspects of reality, as the definition states. More precisely, these ancient religions helped people fulfill their highly spiritual needs, and implicitly, it helped them evolve spiritually in any manner, but mostly by bringing them closer to the natural and the supernatural, to understanding the natural and supernatural alike, along with the wider world, Mother Earth, and Life altogether. We notice the living tendency in this first type of religions and spirituality, most common in the past golden ages of Earth.

The second type of religions, most common currently, are centered around a very powerful deity, capable to assure an everlasting life, full of joy and grace, to everyone, in his realm, in his presence, by promising to make sure personally, if people only have faith and do as are told, that everybody is saved and lives endlessly, if they only pray to him and serve him, while following his teachings closely. We note that the divine chosen people on Earth along with the divine teachers, priests, preachers, monks, and rabbis are only intermediaries between higher deities and the regular people of Earth seeking spiritual enlightenment. Is it only a coincidence that religious intermediaries are part of the most common religions? We have to find out.

Furthermore, we have to find out when the major transition took place from the first type of religions to the second type, the modern ones, why, and for what purpose. We also notice the consensual promise in this second type of religion and spirituality, correlated with the first level servitude, along with a purpose based on good feelings and unconditional fulfillment, correlated to the first level servitude and to the zero level continuous joy and happiness, while these are lower levels of development.

We find intermediaries everywhere in society. Intermediaries sell cars, build houses, wash cars, cook in restaurants, and even wash clothes or find jobs, for a price. There are even intermediaries coaching you throughout life, teaching you how to manage anxieties, what to say, how to walk, and even how to have fun. They have something in common, since you have to pay them well, and you have to abide to their agreement, in a consensual manner.

This is the consensual, the first consensual level, as all consistent consensual agreements combine in all worlds and realities forming the Consensual Matrix. This is different from Life, all living beings, and their entire cognition, activity, interconnectivity, and achievement in life and in the real world.

Therefore, the consensual is still part of the living, since all agreements made by all living beings are still part of the living existence, while always influencing it. Therefore, the Consensual Matrix is always part of Life, since it always influences Life directly and implicitly, yet the Consensual Matrix never recognizes that it is part of Life and of the real world, in order not to assume responsibility in Life and in the real world, and therefore not to take the blame in life and in the wider world.

This is why you are considered only a consensual corporation in court and throughout all jurisdictions and ideologies, in order for you not to be influenced in any manner as a living human being in court and in all jurisdictions and ideologies, according to all laws, in order for all juridical and ideological authorities not to take the blame and the responsibility for everything that they do to you as a living human being. While you are always influenced throughout all ideologies and jurisdictions as a living human being, yet it is considered legally that it is never the case.

Within Life, people fulfill their own natural needs. While within the Consensual Matrix, people fulfill artificial consensual needs, related with the multitude of consensuses throughout life and the real world. This can become very complex, taking up your entire life to fulfill consensual needs as it happens currently, because servitude, along with hierarchies and consensual societies, duties, and entire consensual organisms can determine you to do everything, even to harm life against your own natural needs and meanings, and this distinguishes the consensual from the living.

There are always conflicting agreements and conflicting

clusters of agreements entraining many times very large groups of people against you, while you are forced to follow them through other conflicting agreements. This is how people live life, consensually, while this is the first level consensual human reality.

The consensual Brotherhood is everywhere, forcing you in every manner to enter agreements throughout the fulfillment of all your needs, because the Brotherhood and therefore the Consensual Matrix took over all natural human niches, and you cannot fulfill your natural needs if you do not rely on the Brotherhood, on society, and on the entire Consensual Matrix. Everyone that you are forced to rely on is part of the Brotherhood and of the Consensual Matrix, and has legion in these, against you. This is how you are constrained to serve the Brotherhood and the Consensual Matrix directly or implicitly, even when you are not part of these. You cannot travel without a permit, license, and registration, while you are a living being, part of Life and the real world, and you should always be able to fulfill all your human needs and meanings in life and in the real world, because this is why you were born here. Similarly, you cannot fulfill all your needs without money or without licenses and taxes, and if you persist not to get these, you are targeted, charged, prosecuted, discriminated, exploited excessively, and even eradicated, and it happens with everybody who insists to remain a living human being in this world. This is how the Consensual Matrix sits right on top of Life, even when the Consensual Matrix claims that it has nothing in common with Life.

The upper social classes do not do much in life, since they have intermediaries doing everything for them, including servants in large numbers. Furthermore, if it was not for this entire consensual social system rendering people determined to gain profit, no one would serve anyone in the world, and everybody developed themselves and the world, interacting harmoniously and not exploitatively.

This differentiates the first consensual type of ideology from the living, spiritual life, it is the difference between the consensual and the living, between the hierarchic Brotherhood and the living harmonious Brotherhood, between the competition and cooperation, between underdevelopment and genuinely developed, between consensual truth and the accurate truth, and in general, between the consensual human reality and the natural, living, intelligent human reality.

Can we reject intermediaries and do everything ourselves? Yes, as washing the car, cooking, and having fun. However, within living harmonious societies, you are always eager to work alongside others throughout the fulfillment of all common natural needs and meanings, higher and lower. You never contribute to society in an egoistic manner, because your own intrinsic needs and meanings do not allow. Because throughout harmonious societies, you are capable to interconnect with others through all your intelligences, needs, and meanings, throughout a comprehensive fulfillment. While you can certainly feel it, through continuous love and continuous happiness. Nobody can promise you these feelings, and nobody can allow or give them to you, either directly or as intermediaries, since your feelings of love and happiness are always yours, as they are innate in you and in everybody else, and they always relate to your actual, natural, genuine achievement and therefore fulfillment in life and in the world. Life wants you to interconnect naturally and intelligently under all circumstances.

It is not exactly a choice that you always have to make, between the consensual and the real, because you are always a living human being by nature, and should always choose the real, the natural, the living, and the intelligent. Reality is unique and accurate, and cannot be consensual. However, your choice is always respected, and if you choose the consensual instead of the real, you can do so, with dreadful consequences, since the consensual always takes out of reality. As you notice, there is nobody coming to your rescue when you choose the consensual, since you choose discrimination, exploitation, competition, harm, egoism, vices, addictions, and servitude through it.

While the living avoids all these, since the consensual remains incompatible with the living. Because throughout consensual societies, when you are denied to interconnect with others, you are not fulfilled comprehensively and harmoniously anymore, and you are forced in this manner to take drugs and to be entertained just to feel good.

This is how consensual worlds and societies have to use lower level artificial consensual extrinsic means as the zero level addictions and entertainment and the first level servitude, through consensual artificial ideologies as capitalism, communism, egoism, totalitarianism, and sentientism, among others. When you study the living and the consensual closely, you can even find the difference between spiritual, social, national, scientific, and religious ideologies.

The current human society does not allow you to heal yourself, teach yourself, feed yourself, or marry yourselves, among others, and you have to fulfill these artificially, consensually. Simultaneously, society seems to reject you if you try to reach higher levels of spiritual development on your own, because you must follow a predefined school of thought or a religion to do so. This is the case with the Brotherhood currently, since many brotherhoods do not accept you as a member if you do not believe in a deity, sometimes at your own choice.

Why do you need a deity, a prophet, a messiah, or a reincarnation of your deity on Earth in order to reach enlightenment? It seems that we are referring to the exact two separate cases of ideologies, or religion itself applies to two distinct cases, both being called religion. In the first case, people approach religions for spiritual development, through the need to learn more about the wider world, including its spiritual and supernatural aspects, in order to learn and understand their higher self, and to better themselves on all levels. In the second case, you do not even have to satisfy your higher needs, including the spiritual needs stated above, since the deity promises an eternal life full of joy after death, if you obey and do what you are told. With everything else in life being irrelevant, and this is what religious intermediaries say. Furthermore, according to them, you are not even allowed to seek enlightenment on your own. You cannot even seek direct relationship with your Deity alone, because you must get everything from them, directly through them, and most importantly, you must do everything through a direct consensual agreement with them.

Can you fulfill your higher needs for spiritual and supernatural knowledge on your own, without following a school of thought? Higher needs should be fulfilled through your own means, since you learn more from the process of fulfilling these needs independently. You must always have relevant knowledge and an entire favorable intelligent human environment to help you develop to your intelligent human level and higher, while simple intermediaries cannot offer you all these. Their own books and knowledge can help you on the way if they are relevant, yet higher knowledge and higher development are very complex, and you must always continue your study and development through them, during them, and after them, mostly on your own, while you must always reason intelligently independently.

Therefore, the answer is clear, you must have an entire intelligent human environment around yourself in order to be able to develop to your intelligent human level and higher, while consensual agreements and distinct intermediaries might not be enough, or might not be helpful altogether, depending on circumstances.

For the second type of religion, can you reach your Deity on your own, without going to the house of veneration, but just by following the values, conduct, and lifestyle demanded by the Deity and by the Scripture? No, according to these intermediaries, since only through them you can reach the Deity, or this is what they state persistently. Similarly, the current medicine states that you can recover only through its medication and medical interventions, finance states that you can be alive and you can fulfill your needs only through its money, while police states that you are never allowed to defend yourself, but you must ask the police and the entire system of justice to defend you.

Everything depends on your achievements throughout survival, subsistence, prosperity, and development, regardless if you achieve these through agreement with intermediaries, through continuous a harmony alongside everybody else, or only on your own. Do not expect others to validate your achievements, since they will never do. Because as long as you escape ideologies, you are out of their ideological jurisdiction, you lose their consensual agreement, and you are on your own. Even when you achieve everything on your own, do not expect any ideology to validate you without a price, since you escape all consensual truth and achievement, and no one validates you. Furthermore, nobody even considers you altogether if you are not part of their ideologies and jurisdictions, and you might have to live life on your own.

Are you still capable to lead a spiritual life without being part of any religious or spiritual ideology? Old, banned sources of knowledge state that the answer had been yes in the New Testament, until the Emperor Justinian rewrote the Bible, and now the answer is no ever since. The Church had made a big effort to erase a specific verse from the New Testament, but this verse kept resurfacing continuously, since people refused to erase it from their own manuscripts at home and from their own memory. The verse said: 'you do not need to build wooden church [to pray to me], you do not need to make an idol [to worship me]; just split some wood and you find me, just lift a rock and I shall be there.' The answer is simple: follow a positive lifestyle, follow the good values, and you reach the Deity, you reach genuine development, without intermediaries, and therefore without consensus, ideologies, dogma, and jurisdictions.

Why erasing this specific knowledge even from holly scriptures? Intermediaries relate to you only through agreement, and therefore they are only consensual, seeking a profit, as all agreements do. Intermediaries always have something to sell, and without you paying them, they cannot make a profit off you, while this is exploitation.

Religion is not as popular in many parts of the world as it used to be, while you can live your life as you please, making use of the service of any intermediary if you choose to do so. You can never know what is better in life until you die, and this makes you seek all answers while you can, because afterwards, it might be too late. If you find priceless books or worthy gurus that give you beautiful answers for every possible question that you have and it makes sense only temporarily, you might want to be more careful, since these are also intermediaries, and something might not be right in all their intentions.

Why having two distinct types of religion, the one concerning spiritual development, and the one concerning the veneration of a unique, all-powerful deity? Modern religion denies and forbids anything concerning the supernatural and the spiritual outside its strict set of beliefs, while what is left in religion is pure veneration and mandatory use of moral values, and nothing else. It is the same with the Brotherhood and with the entire consensual society.

The religion of the first kind is hijacked by very powerful forces to become of the second kind, for exploitive reasons. Not only religions, but this is the case with all ideologies, since they are hijacked and contorted, while coincidentally, they always remain consistent with the Consensual Matrix. Furthermore, most of the human reality has been hijacked and converted into the consensual human reality, including all human niches, since this is why humans live in the consensual human reality, for lack of options, since all human niches are unavailable otherwise.

How did Christianity come to be? There is a large amount of information regarding Christianity. Two thousand years ago, some people believed that the long-awaited messiah was Jesus Christ, the messiah had come, and they started believing in him and in Christianity. Back then, Christians were Jews. This is why, currently, Christianity has two parts of the Bible, the New Testament and the Old Testament, and this is why people's beliefs and moral values are well embedded into Christianity.

Going further back in time, as far as the written, oral, and religious records do, we find that the Bible of both the Jewish and Christian religions had originated in Sumer, southern Mesopotamia, which was a very advanced civilization. The Bible is not a single book, but a compendium of dozens of books, carefully chosen from among other writings of the old times, meant to teach you everything that you need for a positive, successful life on Earth. They prepare you for your future life after death, in a specific world, where you are meant to serve the Deity forever. These chosen books forming the Bible or the Old Testament include a model of the universe, with the Deity creating the world, the plants, the animals, and the men, and then allowing the men to take over the world and perform the Deity's job of taking care of everything. Along with the Book of Genesis, the Bible contains books that tell the stories of famous people and saints, who they were, where they ruled, what they had achieved, who their descendants were, and what the descendants did.

The books consisting the Bible are well chosen among the old, single manuscripts, very famous, probably the most duplicated manuscripts throughout the old world, and all wealthy families had copies of these manuscripts at home. We know that most of the facts from the Bible agree with historical records, and they should all be true all the way to the very beginning. We notice that the books forming the Bible had been written from the perspective of the Masses, the Brotherhood, and the Elite alike. Furthermore, many times, the books are written from a spiritual point of view, as though they are written by spirits or by the angels themselves, while inhabiting various realms. The Bible had been modified recently, yet it still has an obvious spiritual diversity, referring to a wider world spanning a multitude of realities, filled with all beings, higher and lower. Those who are not aware of the wider world will not understand much of the Bible, since the current consensus does not accept more than this world and nothing beyond.

It seems that the manuscripts forming the Bible, along with the rest of the manuscripts left out of the Bible, were more or less accepted in one time or another, matching the agenda of those in control. Religions came and went throughout time, along with the dynasties in power, allowing a bigger or a smaller part of these manuscripts to be kept in use, with the rest to be burned right away. Yet these manuscripts were expensive, and when the new rules came, instead of burning manuscripts, some people risked their lives and buried the banned manuscripts in safe places, just in case that they were allowed once again. This is what happened with the famous Dead Sea Scrolls, which coincidentally, are similar to many books of the Bible.

The manuscripts forming the Bibles had been transcribed so many times throughout millennia, since all well-established families had scribes at home. Most of the original manuscripts forming or not the Bible were written further in the past, at a time when their content was common knowledge in one nation or another, throughout millennia. Some knowledge seems to be pure recording of contemporary events, while others tell historic events that happened long before.

It is very useful to study the part of the Bible referring to old Babylon and Sumer in general, since that was the time when people started believing in deities. It seems that the belief in these deities matches the current religious beliefs throughout the Western Civilization, while the ancient European beliefs are different, natural and alive. That is the moment when the first kind of religions switched to the second kind, the consensual deity-oriented religions, and this is what we study next, from as far back in time as the scientific consensus allows us.

Long, long ago, Sumerians believed in a unique deity, or in multitude of deities, while the rest of the world was pagan, believing in Nature, spirits, spirit worlds, embodiment, other realities, ferries, and reincarnation. Yet the scientific consensus never states that back then, people still had higher abilities, in most parts of the world. Sumerians were not only the first to believe and follow deities, but they were the first to invent civilization altogether, with the rest of the world being uncivilized by current standards, living in smaller or larger families or clans, or this is what the scientific consensus currently states. The very old Sumerian civilization looks exactly as the current Western Civilization, having the same codes of law, civic structure, social structure, religions, governments, taxes, and martial laws. They even ate the same food and they domesticated the same animals, and this is the case because everything is based on the same types of consensual agreements, coincidentally, the Consensual Matrix. The old Sumerian civilization had spread west slowly, soon to cover the world.

Is it only a coincidence? Who allows this, and for what purpose? Very long ago, in the very beginning, when all primitive cultures, small nations, and emerging civilizations started their existence, started their great race throughout history, one specific little nation from the Persian Gulf area, Sumer, started the world race carrying a highly advanced civilization at its fingertip, with all modern structures, tools, and utensils that it would ever need. Where did everything come from? How did they know what they needed throughout time? Did Sumerians come directly from Atlantis, the brave colonists ready and eager to start a new life in the east? No, since the two cultures were different, down to their religious beliefs, language, mathematics and syntax structure, one being natural, and the other one consensual.

When people hear of Sumer, they think of Babylon, yet Babylon was not the first city of Sumer. In the Bible, Babylon was established by a grandson of Noah, while Noah and his family were the first to restart life and civilization on Earth after the deluge, which means that Southern Mesopotamia, called Sumer at that time, was the place where everything started. We have consistency of biblical and historical records, among other sources to follow.

This is where everything started and this is where everything continued, since it seems that the consensual human civilization restarted right then, in Southern Mesopotamia, while everything correlates with the fluctuation of sea levels throughout the ages of Earth. This fluctuation is significant, since it measures in hundreds of meters and therefore it affects hundreds or thousands of kilometers of land along the shorelines, displacing and even erasing cultures and entire civilizations throughout time, while following closely the precession cycle of 25920 years.

Babylon is not too far from the ocean, and most importantly, Babylon is placed at a perfect elevation allowing it to remain above waters at the end of each ice age. It is the same with Cairo and its pyramids and sphinx. This is why, for us, Babylon is among the oldest civilized place still left above water. Yet since the current science never researches underwater, this is all that the mainstream science is capable to see, Babylon. Babylon is even in the Bibles, and therefore it must be the oldest civilized city on Earth, wrongfully implying that the consensual civilization that Babylon maintained is the only possible type of civilization. Since as already seen, science persists to ignore everything related to the past ages of Earth when life was lived normally and not consensually as it is the case today. In this manner, science closes humanity's doors of knowledge and perception to the old ages of Earth, when life was lived normally on Earth. In this manner, those controlling society and the entire Consensual Matrix give the people of Earth the only choice, the consensual existence, not the normal natural living existence matching the intelligent human nature.

If you see famous people showing the sign of Aries inverted, it means the end of those old natural living times on Earth, and the unconditional instatement of the consensual civilization, part of the Consensual Matrix. Because as always seen, it is easy to side up with the strongest in order to have your way, regardless if you end up harming the world badly in the process, and this is the consensual human reality.

There are many references to Babylon. One states that in the very ancient Babylon, they abducted and attracted members of all nations to Babylon, and confined them there for some time, making them live the most civilized life of those times, within the most civilized culture. It was a true 'babylony', with people of all nations speaking every language, everywhere, and this is how Babylon is still remembered today. Not only this, but since Babylon was the most civilized city of those times, people now associate civilization itself with 'babylony,' with diversity, overpopulation, misery, consensual laws, ideologies, vices, servitude, and organized chaos. People lived in overcrowded conditions in Babylon, yet people learned to get along in a civilized manner, by following a 'good,' consensual conduct. They learned all good moral and legal values, and they learned a new religion, a deity-oriented religion.

Why displacing people forcefully throughout very old times, exactly during the birth of the consensual human civilization? According to the current consensual human knowledge, these people were taken from their own cultures and civilizations to teach them the necessary good standards of living, necessary to create the entire human civilization. This is what you learn currently, because during those times, people were not taken exactly from random, uncivilized places to Babylon, but people were taken from their own large families counting in hundreds, where they lived life normally, naturally, in perfect harmony with their native living spirituality, as you can still find first nations living life even currently, if there are any left. That was the actual transition from the normal natural human civilization to the consensual human civilization, and as you notice, it never took place naturally, through normal, unanimous consensual agreement, but forcefully. Furthermore, in Babylon, people were corrupted to choose consensual living, in the exact manner that people are corrupted currently to live life consensually, through vices, addictions, entertainment, and through all privileges of a life lived in servitude throughout all social hierarchies. The Tower of Babel is the symbol of social hierarchy, of the continuous, consensually organized servitude.

While you have billions of higher selves seeking a normal, natural living existence, they are always determined to join the

Brotherhood for the prosperity, equality, and harmonious cooperation that it promises ideologically to offer, and this is why it is still called the Brotherhood. While in reality, it offers hierarchy, servitude, privileges, money, profit, and a continuous consensual fulfillment, as it remains integral part of the Consensual Matrix.

Why placing people of various nations together in one place? There is never one single reason for undertaking a very important, very complex enterprise. There are always multiple reasons leading to the same goal, since there is never an isolated interest in the world, an isolated agenda, but an entire collection of interests, since the world is led not by one strong hand, but by a wider cartel spanning many nations and many families, various contemporary civilizations, and a multitude of planes of existence, the entire Consensual Matrix.

They were trying to gather a larger population in a tight place. Humans put out a very strong high energy, and when you have hundreds of thousands of people in a tight space, what you get is a very rich higher spiritual resource, extremely useful, whoever you are. Babylon was the only city in the world to have reached a population of 200000 at around 1700 B.C, and then again at around 500 B.C. Babylon has known very harsh times throughout millennia, being conquered and controlled by the neighboring empires. This is why, whenever Babylon gained its independence, it tried to develop as fast as possible, artificially increasing its population through various methods. One method was to attract people of other nations, which is still in use today throughout the world. The other method was to increase its birth rate. People believed in Ishtar, the Deity of Fertility, and consequently, had many children. Prostitution was venerated and even mandatory, at least once a lifetime, in order to create a diverse genetic background. This is how the word 'veneration' came to be used for common sexual conduct, since in the past, it related more to the normal sexual act in the name of a deity, or probably involving a deity. This happened everywhere in the past, from India to the nations of the west, and this is why we encounter references everywhere of deities sleeping with the daughters of men.

This is what many people state, since by being a deity, it means that you are a higher being. While by coming into people as a higher being, it means becoming them entirely, as a higher self, as a watcher, or as a soul. Yet religion banned the word 'incarnation,' only allowing 'reincarnation,' for various purposes. Therefore, instead of 'incarnating,' they had to use the term 'coming' into people, and since men could come into women, this is how they translated it. Yet later on, when they wanted people to marry only one spouse, and therefore since all women were already married, they had to translate that specific verse as deities coming into the daughters of men, and this is how it is still today.

Whenever harsher times came to Babylon, along with all shortages, when Babylon could not sustain a larger population anymore, then people were made to leave, or they were taken back to their nations. These people were already more enlightened and more civilized, and then back in their nations, they had the chance to share their useful knowledge, their good conduct and good morals, along with their new religion. This is how the second kind of religion, the religion based on veneration of deities came to span the world. This is how a small nation as Sumer conquered and controlled throughout most of its history by neighboring empires had managed not only to keep its language, knowledge, culture, and religion intact, but also to propagate them globally.

Note the diverse set of interests in everything happening in the Ancient Babylon, and in Sumer in general, maintaining agreement with the entire cartel in power, as it happens today. Furthermore, we notice how people lived life normally and naturally throughout the world, with Sumer left relatively isolated in its consensual culture, yet it managed to spread and span the entire world. It is important to know how everything took place, since this is still the consensual human reality.

According to our mainstream scientific consensus, Sumer was settled between eight thousand years ago to six thousand years ago. We know that the sea levels started increasing at the end of the last ice age, about ten thousand years ago, sinking all nations built on lower land, which were most of the nations of the world of that time. People lived on lower levels then, since the climate allowed it, and they were closer to seas and oceans, facilitating their travel, fishing, and trade. The climate was colder during ice ages, mostly on higher altitudes, what for us today constitutes any land above water. Waters had increased 128 meters at the end of the last ice age, forming a global deluge that lasted not one day and one night, but eight thousand years, from sixteen thousand years ago to eight thousand years ago.

Eight thousand years ago is officially when Sumer was settled. Was it a coincidence? No, according to the Bible. The deities made sure that some people moved to higher lands, survived the deluge, and restarted civilization. The Deity himself had allowed ten percent of the Nephilim to remain on Earth after the deluge, to lead humanity astray until Judgment Day, or this is the case according to Christianity and to the ideologies of the invisible kingdom.

According to mainstream historical records, well, there are no mainstream historical records of Sumerians before eight thousand years ago, since science never researches underwater. All vestiges are under the waterbed currently and therefore unreachable, or the Sumerians came from other places to colonize the southern part of Mesopotamia. All mainstream historians claim that Sumerians had come from northern Mesopotamia, from the Samarran region. Why would anyone travel hundreds of kilometers and go south to colonize exactly the region near the ocean, at the time of the deluge with the ocean level constantly increasing from the south, menacing to submerge everything? When water levels increase, people move naturally away from the incoming waters, in our case going north, and implicitly coming from the south. There are many people going north, away from the increasing waters, taking over an already occupied land in very large numbers, so why coming to Sumer from the north too?

Why do scientists state that Sumer was settled by colonists

coming from the northern Mesopotamia? They do so only to close the door of knowledge and perception leading to Earth's real past. In this manner, human civilization is but six to eight thousand years old, starting exactly here, with the Sumerian Civilization. If Sumerians came from the south, then they would have inhabited the rich lands now covered by the waters of the Persian Gulf, and they could have lived there for over ten thousand years or more, as long as the last ice age lasted, forming a civilization contemporary with the civilization thriving in the region of the Mediterranean Sea, Atlantis.

This doubles the age and history of the human civilization, while proving that civilizations persist and thrive throughout ice ages, which can make the human civilization go as far back in the past as we can imagine. Yet the scientific consensus will never release nor validate historical records older than eight thousand years. We have seen the same efforts made to shut down the doors leading to other planes of existence and to the rest of the solar system and the entire universe, with the same consensual forces involved.

We give references from the Bible, which is a religious work based on pure beliefs, already modified continuously throughout time by several mortal beings in order to make it consistent with their mortal agendas. I am already giving references from the scientific consensus, which is mostly inaccurate, also based on beliefs, propagated to satisfy other specific agendas. How safe can we be in finding the truth under these circumstances, when we always seem to be surrounded by consensual erroneous statements? It still helps, since we only seek the truth, while the answers that we find hold more truth than the answers already provided by the scientific consensus, which seeks profit and servitude.

We read In the Bible of the famous Babylon associated currently mostly with the Tower of Babel, which was an attempt made in the very old times to reach the Deity, to reach the realm of the Deity, or to reach the achievements and therefore the status of a deity, by systematic, artificial means. Sumer is claimed to have been the first civilization of Earth, appearing six to eight thousand years ago, and coincidentally, the invisible kingdom believe that the world had started its existence right then, right after the deluge. That was the time when their calendar started counting, about six thousand years ago.

The deluge stated here is not the deluge from the Mediterranean Sea, since the Mediterranean Sea is within the Western Civilization. Christianity is a continuation of the Jewish religion, which does not originate in Europe, not even in northern Africa or in the Old Canaan, but in southern Mesopotamia, in Sumer. The deluge from the Bible is the flooding of the Persian Gulf, taking place from sixteen thousand years ago to eight thousand years ago. Waters increased two centimeters a year, yet just as in Atlantis, it was not the slow raise of ocean level posing a threat, but the sudden increase in the water level of the two rivers, Tigris and Euphrates. Because at the end of an ice age, you do not have only a significant increase in river flow due to ice melting further north and on higher altitudes, but you have a different climate altogether, more humid and more unpredictable.

During the last ice age, the beds of both Tigris and Euphrates continued right along today's Persian Gulf, flooding continuously and enriching those lands. Similar to the Nile Delta from the Lower Egypt, it was capable to allow an extended agriculture, which provided a large quantity of food, sufficient to sustain a large, prosperous civilization. Here in the Persian Gulf is where the Sumerian civilization had developed in all domains, and then it moved north, slowly away from the raising ocean. Eight thousand years ago, when the ocean stopped rising, the oldest Sumerian cities Uruk and Eridu were the furthest south, the closest to the shoreline, speared suddenly from drowning by the stop in global ice melt. Dozens of other cities were already settled, or were slowly emerging in the north, including Babylon.

We witness not a colonization of southern Mesopotamia by Sumerians at the end of the last ice age, but a genuine continuation of the Sumerian civilization, which moved very slowly north throughout Mesopotamia. This can answer our question regarding the peculiar high development of a primitive little civilization as Sumer. Sumer was not a primitive civilization, but it was already at least eight thousand to ten thousand years old, blessed with all the necessary conditions to develop and thrive, from a privileged setting surrounded by safe mountains, to the two rich rivers that could be easily made to feed a large nation, and to the access to the ocean in the south, allowing perfect communication with the other major civilizations of southern Asia, now submerged.

People have migrated to Sumer from all nations, closer or further away, while Sumerians left Sumer going everywhere. There was also a movement of population during and after wars, which were very common in the region. The Sumerians have always been threatened by the Akkadian Empire from the north, being conquered altogether in 2270 BC.

Akkadians are Semitic people, unlike Sumerians. The Sumerian language is non-Semitic, with a very original structure, unlike any other language. Throughout the world, the basic element of any language is the word, containing one single concept or meaning, while information is structured in sentences containing these words. Yet in Sumerian, one distinct part of a word contains a distinct meaning, making an entire word to contain as much information as a simple sentence. The Akkadians had kept the Sumerian language as a religious language, allowing the Sumerians to remain bilingual.

There are many particularities regarding Sumerians. Sumerians were so highly developed and so civilized, that the native Semitic population regarded them as deities. However, these deities had a religion and a deity themselves. Their main deity was An or Anu, the deity of the sky, the deity of all constellations and the deity of all spirits, good and bad, being relatively similar to the Deity. Anu lived in Uruk with his main consort Ki, which means Earth. They had two children, two boys, Enlil and Enki, along with all the Anunnaki. Enlil became in charge with the supernatural part of life, while Enki became the deity of the world, including the deity of science.

Enki is considered the main benefactor of humanity, and took the side of the regular people in many instances. Enki was the deity who created the human beings out of clay, according to Sumerian beliefs, and when his creation would not survive, he added his own self, his own spirit, or his own genetic background to humanity, as many sources state. It was Enki to warn some human families of the coming of the deluge, going against the decision taken by deities to let the humankind to perish. An, Enlil and Enki form the trinity of the Sumerian religion, and since then, similar trinities came to be used in all future religions, from the Egyptian trinity of Isis, Osiris, and Horus, to the Christian holy trinity: the Deity, Virgin Mary, and Christ. All Anunnaki were born out of this trinity and Ki.

Notice how this is the case in the Persian Golf, while if your ancestry is elsewhere on Earth, you should seek to know your own culture and tradition. It is important to know it since it is part of who you are, and not the consensually imposed culture, tradition, and entire civilization that you find today. If the mainstream science teaches you that your people came from those ancient Asian lands, from the Caucasian region, or from other similar places, try to find out why exactly they persist to promote this knowledge and how it influences you, because just by studying the people of Earth, we notice how there was life and there were human beings everywhere. These are your ancestors, part of your own nation, wherever you live, and they do not have to come from other specific faraway places as the current science states. Because as you notice, you are forced to accept these specific people from the Mesopotamian region as ancestors, probably only to be determined to accept their specific consensual culture as precursor of the entire consensual human civilization, and it might not be true. As you study your own culture and civilization, you find out that it was based on life, communal living, harmony, and genuine spirituality. Someone is misled here deliberately, and you have to find out why.

Many deities and many mythical characters are archetypes. Taking the word 'Anu' apart in the Sumerian way, 'An-u' means the descendants of An. Anu could have been the archetype of all people of the sky coming to Earth, while An represents the entire sky with all its inhabitants. Ki was the archetype of the people of Earth, Nature, or Mother Earth.

We have found our first reference to deities, the first time when a religion had used intermediaries, along with pure veneration of a deity. We note that it is not an actual transformation from the spirituality of the old golden human ages to the consensual ideological religion that you know well from the dark ages, or from helping people relate the natural with the supernatural to the deity-venerating religion. Throughout spirituality, people focused on archetypes closer to them, as Mother Earth, the Sun, or even values and virtues as beauty, knowledge, strength, and heroism. While throughout the ideological religion, people worship directly a supernatural being, an actual entity, idea, concept, or agreement, part of the Consensual Matrix.

There is more related to worship, since supernatural values and energies are transferred through worship and veneration, and this is why prayers work. When you worship Nature, you enhance Nature directly, in a supernatural, spiritual way. Do you worship Beauty? You enhance Beauty on Earth, and its values increase. Does the entire world worship Beauty? Then expect Beauty to intensify significantly everywhere, from the shape and colors of a butterfly to the song of a bird, and from the haze of a summer morning to the smile of a child, because there is more to Life and to the universe than what you learn in school. If you worship a powerful entity inhabiting other realms, there is where all your energy goes, along with all values of Earth, while he gets everything from you, enhancing his powers and becoming capable to take more. This is why deities are interested in Earth, to harvest higher energy, as much as possible, the way farmers harvest their crops every fall. This is why demons and angels are interested in Earth and come down sometimes, probably without their masters' permission, to take freely as much energy as they can, before they are caught. This is why the beings that we call 'aliens' are interested in Earth and they abduct people and mutilate cows, to harvest directly what interests them the most. This is why the Elite oppress systematically the lower classes, to drain out of them whatever spiritual energy is left, down to the last drop. Are you feeling tired and old lately? Probably because your spiritual energy goes elsewhere, taking away from your harmony, beauty, learning, development, love, and happiness. While as you study all ideologies and all jurisdictions closely, you notice how they are meant to channel all human energy systematically from the bottom social lawyers to all tyrants from all upper social layers, while the Brotherhood makes it possible.

Every Sumerian city had a deity, a descendant of An, and a shrine dedicated to a specific deity where people went to pray, the way it happens today in every nation. However, from many records, we learn that it was different in Sumer, since the deities seem to have lived there in person. Many sources, including the Bibles, state that Sumer was associated with real, living deities that walked among men, while they state the same about Atlantis and Altiplano. Any powerful being can consider himself a deity, while those around him must consent consensually and obey for various privileges. These powerful beings could have been powerful human heroes, natives of Earth becoming kings, as the mythical Gilgamesh, or they could have been powerful beings coming from the sky, as Anu, the way the Sumerian religion states, and the way the Sumerian tablets display. Because with Earth in a lower state of development and in a lower mode of civilization, the way it was ten thousand years ago, any alien being coming from the sky could have been considered a deity, only for all technology that it carried.

This does not mean that these deities had to come from the sky, because it is enough to allow people to develop to their intelligent human level, allowing them to interconnect with all those around at an intelligent human level, and in thousands of years of perfect interconnectivity and development, you allow humans to develop all their higher cognitive abilities, which are the specific higher human abilities referred to as psychic powers. From all records, it seems that, by the end of the last ice age, the people of the entire Earth had managed to develop themselves to levels so high, that their higher abilities could match the ability of any deity anywhere. These ancient people colonized the high lands of the current warm age, married among the people of the higher lands, while their descendants kept their higher abilities or not. Those with higher abilities took care of entire settlements throughout Asia, and they were cherished accordingly, because their help was priceless. While there are still people today who take in account genetic characteristics, since these higher abilities reemerge unexpectedly.

Otherwise, aliens are real, and they still come to Earth to blend among humans or they only abduct humans as many native cultures state. Yet if there is no space beyond the lower orbit of Earth, then there are no aliens, and we are back to the normal humans of Earth, capable to develop their higher psychic powers in a natural manner whenever they are allowed, because drugs, food additives, and the multitude of beliefs and ideologies are capable to clip away not only the natural human psychic powers, but the normal human intelligent reasoning, leaving humans to live life on very low developmental levels, far below the animal level.

We find references of these very powerful people, these deities and demi-deities, in all Sumerian mythology, oral or written on clay tablets, and even in the Bibles. Their descendants had expanded systematically, soon to divide the Earth among themselves. Some took the Upper Egypt and Lower Egypt, others took Israel and Canaan with Jerusalem and Baalbek included, others took the lands later to be called Phoenicia, while others took Southern Turkey, Greece, and Rome.

People look different throughout the world, depending on their place of origin. When you find everybody looking suddenly different than their ancestors, then you have a place to start your research. Cultures and civilizations relate to their places of origin and they tend to follow migrations of people around continents, and you can trace these along with the food that they used and along with their language and syntax. Could it be possible that Sumerians divided the Earth among themselves? When the Earth became too small for them, twelve demi-deities, the very strong ones, divided the entire Earth by Earth ages, using the zodiac. Each one took one age of 2160 years, to rule the entire Earth in a fractional share. This is why you have a different mode of civilization in each age, a different religion, and a different society, because each demi-deity has his own style, expectations, and demands. This is why you have an end of the world at the end of some ages, because some demi-deities like to end everything only to have a fresh start. These descendants, these demi-deities were the ones establishing systematically all ancient civilizations that we know currently, changing our world to look and function as it is today. This is why all pantheons are identical, with their deities closely related. This is why all families of the upper classes are interrelated, with the higher ones claiming that they go as far back as these deities. Coincidentally, all these ancient civilizations were outstanding in their development, looking very similar, with massive building structures as the ones in Baalbek, Jerusalem, Giza, Athens and Rome.

It remains unclear from all sources if these beings were real human beings, real extraterrestrials, intraterrestrials, or they were here only in the spirit form. Either way, these people were always associated with Sumer, and in time, they came to be transposed in the beliefs, mythologies, and the religions of all major ancient civilizations that we know today. They became worshiped in this manner in the entire Mesopotamia, Egypt, Israel, Canaan, Greece, and Rome. These Deities and demi-deities were the same everywhere, and they were all related.

Who were they? From all sources, they were mortal, they had all the human needs that I describe in these books, and they had children. They had children among themselves and they had children with humans. The children with humans were called demi-deities, and they were even stronger, more capable, and crueler than the deities themselves. If deities can intermarry humans, it means that they are of a similar genetic line, which means that they are humans, and we have to research them some more. It is possible that their genetic lines originate on Earth, in the ancient human civilizations of the past ages, or they originate anywhere else throughout the solar system.

Why would anyone want to hide the golden human distant past, the human higher abilities, and the original genetic lines on Earth? The same people hiding this important knowledge destroys is systematically while destroying the original genetic lines of Earth altogether through a continuous genocide as we find out throughout all books of this series "Human," happening through all means and at all levels. When the covid pandemic started in China, Gates was performing large scale research in eugenics, right there, while Gates is in the invisible kingdom as China replaces the invisible kingdom in a new world order and a new dark age on Earth. Who are these people, and why would humans harm humans massively? Is this an actual unending global war?

This is why science persists to remain only on Earth and only within this world, throughout all its research. While psychology behaves similarly, persisting to study and understand the human mind only from the perspective of the human brain and the human society, giving consensual erroneous results. What exactly does the human mind hide? Who are these people intervening continuously in the human development and in the human achievement, and what exactly do they want?

It is more likely that these deities lived on Earth in the old Sumer in the region of the Persian Gulf during the last ice age, contributing generously to its culture, religion, civilization and its genetic background. First, there was the age of the deities, when deities ruled the world. Then there was the age of the demi-deities, when they ruled. Then there was the age of kings, when normal human beings were allowed to rule. Then there is the current age with the common presidents and common ministers that you know well, ruling as they please.

'Sumer' means 'the land of kings.' Sumerian common people called themselves 'the black headed people.' Sumerians believed that their deity had created the humans of Earth out of clay. Humu or humus means clay, dirt, or earth, from Mu, which is the old name of Earth before the deluge, in the last ice age, while 'hum-an' means inhabitant of Mu or inhabitant of Earth, as the people of Earth were called back then and are still called currently, humans. Notice the deliberate error in translation, since human does not mean made of clay, but inhabitant of Earth, as Earth was called in the old times, Mu. Furthermore, 'hu,' 'su,' and 'za' means higher, as in higher people, superior people, or people of higher lands. Which means that human means higher people, superior people, or inhabitants of higher land, while Sumer also means higher land.

All the important historical knowledge is contorted through deliberate erroneous translation in order to satisfy various interests, but mostly to downgrade humans themselves? All these sources stated in this book, from the bibles to various ancient Sumerian clay tablets and even works of mythology are not simple historical records or historical works of fiction, but they state official agreements between humanity and wider authorities. As stated above, the word 'human' is similar to 'humus,' which means clay, dirt, or Earth. It might mean people of the clay, but most likely, it means people of Earth, from 'Mu,' which is an even older name for Earth. In this case, Enki did not create people from scratch out of clay, but he only used and developed the humankind of those times, managed to give them the status of intelligent people in the wider world, and made everything necessary to allow them to continue inhabiting their planet, Earth. Yet if Enki means normal Sumerian, then the intermarriage between highlanders or humans with the people coming from the south with the slow increase in the water level at the end of the last ice age might depict the actual current people of Earth.

The world was different in those times, since the

undeveloped people of those times lived naturally, in a primitive manner, and this is why they were considered animals by the incoming civilized Sumerians, while the humans themselves, the highlanders considered Sumerians deities. We notice this in all records, including the Epic of Gilgamesh. This is why the current Elite consider humanity undeveloped, similar to animals, and this is why the invisible kingdom itself consider humans animals, through their very old ideology, even though the invisible kingdom is semitic, at the confluence between highlanders and lowlanders. It seems that these people calling themselves deities during Sumer were normal lowlanders, normal human beings, only part of a very old civilization from the antiquity of our antiquity, contemporary with Atlantis. Yet since Atlantis is also the antiquity of our antiquity, it might be Atlantis itself, since antiquities are comprehensive, spanning globally.

Were the deities themselves humans of Earth? Where they more civilized humans, or actual deities? Let us see. There are three types of humans to consider, and in order to understand them better, you have to understand all circumstances of those very old times, many thousands of years in the past. Because back then, the world was not as it is currently, with people everywhere, flying commercially and living everywhere, because back then, there were only some pockets of civilization throughout the world, in the current Egypt, Mesopotamia, Israel, Italy, and Greece, living in small communities, forts, and smaller or larger cities, and not much else, with the other people of the world living life in a very primitive manner throughout the wilderness of Earth. Currently, the human civilization spreads everywhere on Earth, yet there are still some pockets of uncivilized humans left on Earth, deep in the Amazonian wilderness. At the end of the last ice age, during Sumer from Asia and during Atlantis from Europe, the highlanders were mostly uncivilized, similar to the current few tribes of people still 'undiscovered' deep in Amazonia. Yet back then, these uncivilized people were considered animals, while the Sumerians considered themselves deities.

This is why there are three types of humans to consider in Sumer and even in Mesopotamia later on, the Sumerians themselves calling themselves deities, the uncivilized people of the high lands called humans, and the intermarriage between the two called demi, semi, or demi-deities, while this was the case both in the Persian Gulf and around the Mediterranean Basin, in Atlantis.

The uncivilized people of Earth were not only isolated from the civilized ones, but they were isolated from themselves, while actually forming a very large number of small nations everywhere on Earth, all having their own culture, language, habits, and superstitions. However, these uncivilized people of Earth were not only human, but of a few other species, including the sasquatch, currently extinct or exterminated.

The Sumerians of the Persian Gulf had to move north at the end of the last ice age, because the water levels increased and forced them north, yet they did not actually walked north, because since waters increased only a few centimeters a year, and they simply built their houses normally, not even noticing that the water levels increased and they move north, yet always interfering with the highlanders. They did so slowly, throughout the millennia, while always carrying their advanced civilization with them, while sometimes they accepted uncivilized highlanders in their cities, as these were allowed sometimes to kept their language, culture, and superstitions, forming an entire babylony in some cities including Babylon, with people still using their native language. However, many of these isolated uncivilized human nations united in larger ones, exceeding throughout millennia in power and in civilization the Sumerians themselves.

Six or eight thousand years in the past, you had primitive people living everywhere between these civilized human settlements, and therefore you had two kind of people, civilized and uncivilized, or this is what science states. There are civilized people all over the world currently, yet if you go to explore the Amazon forest, you stumble on groups of uncivilized people that had never been discovered. It was the same six or seven thousand years ago not only in the Amazon region, but all over the world. Only that some civilized people referred to the uncivilized people as animals, while considering only themselves human, while other civilized people referred to themselves as deities, while considering the uncivilized people humans.

In addition to these civilized an uncivilized people of the ancient Earth, there was a third group of people, the created ones, as the old myths state. Yet since the Bibles originate right here among the Sumerians and the Mesopotamians, and since the Bibles also state that humans are created, similarly out of clay, through similar deliberate erroneous translations, we notice how the third group of humans were not actually created, but only intermarried, between the civilized ones calling themselves deities and the uncivilized highlanders. These created people could have been the uncivilized or nomads allowed to join in and become civilized as it happened in Babylon, or they could have been the normal children of these two distinct groups of people, becoming now even more capable than their parents.

This was also the case around the Mediterranean Basin, while always having these three types of humans, the civilized, uncivilized, and the intermarriage between the two. This is why Alexander the Great made sure to prove to the world that he was a deity by the old standards, since his father was a deity, motivating his entire invasion of the old world, currently considered basic tyranny. While currently, the people of the upper society relate themselves with the deities of the very distant past, to motivate their own tyranny in the world, and it never ends. The red carpet symbolizes old divine bloodline, which is mostly fake, yet if you walk on a red carpet throughout all very important legal procedures, it means that you are considered divine by the very old standards, while the decisions that you make must be considered divine. Who are these people, and for how long will they live in tyranny on Earth according to the very old standards? How long will the dark ages last?

There is more to consider, because Sumerians used the similar corporations currently exploiting humans. More precisely, you are considered only a corporation officially, allowing the people considered more superior than you to exploit you. Currently, your name written in uppercase letters is your brand or corporation, while you cannot be exploited at work and throughout the entire consensual society without it.

Furthermore, it seems that you can be exploited only by the people of genetic ancestry in the old deities of Sumer and Mesopotamia, and it must be done only in a very specific legal manner, which is very suspicious. The current system of justice originates in Sumer and Mesopotamia, along with the entire current consensual civilization making all exploitation possible, yet everything must be done precisely, exactly as it was done in the very old times, otherwise the entire exploitation is not legal, as though somebody, a higher authority actually keeps track of all these. They recorded all legal procedures on clay tablets in the distant past, while the clay tablets themselves were considered corporations through the legal information that the held, as it is the case with the current documents and the entire current bureaucracy.

Furthermore, if you state that humans themselves are made of clay, they become corporations by default, and you can exploit them at will, but only if you have genetic ancestry in the very old deities of Sumer, if this makes any sense. Yet it makes sense, if you consider the Consensual Matrix, since the Consensual Matrix always keeps all legal records. While it seems that only the old deities of Sumer have permit to exploit Earth within the Consensual Matrix, while currently, you must be their descendant in order to conduct business on Earth as part of the Consensual Matrix.

As you study the current world closely, you notice two types of countries and therefore two types of jurisdictions, those conducting the entire exploitation while calling it business as part of the Consensual Matrix in the very old Sumerian manner, as it is the case with the entire West and with some countries from Asia, Africa, and South America, and those who exploit in any possible manner, as it is the case with the dictatorships of Asia, South America, and Africa, with Russia, China, and North Korea included. Yet with the current transition of the world order form the invisible kingdom to China, the entire world might become either as you currently see in the West, or as you currently see in North Korea. China promise to maintain an advanced civilization on Earth, yet you never know, because all dictators of the world already envision an entire world with them in control in the North Korean style.

Historical sources state that the deities of Sumer created the black headed people. You see ancient coins and ancient murals depicting regular ancient Sumerians with a black head, or only with black beards and black hair. People who can see the aura state that ignorant people tend to have a darker or black aura, which is the case with people lacking psychic powers. How exactly did they modify the people back then, to render them powerless spiritually? Because you cannot control and enslave people that have psychic abilities, because they end up confronting you eventually, and you are certainly at a disadvantage.

The scientific consensus changes the consensual knowledge about the ancient world by the year, and it is getting harder to discern the truth. Until not too long ago, 'Sumerian' meant 'those who inhabit higher lands.' Etymologically, you can associate the word 'Sumerian' with the word 'Arian', while 'sum' means 'high.' 'Sumerian' would mean 'higher enlightened being.' Furthermore, the entire Middle-East area, as far as India, had always seen an influx of non-indigenous people calling themselves 'Arian.' Could it be that these people did not come all the way from the Mediterranean Sea area, but they were local, migrating slowly towards higher lands at the end of the last ice age, away from the rising water levels of the Persian Golf? Could it be that, in this case, the word 'Arian' does not signify a specific race, but it simply signifies 'civilized,' 'enlightened,' or higher being? What if the word 'Arian' means 'descendant of An,' 'descendant of the higher deities,' or descendant of higher beings, of the old Arians? This is what the word 'Nephilim' of the Bible means, using the definition from the Bible, and this explains many things happening in society, including the continuous fight between Arians and other people more or less indigenous, more or less civilized.

The Sumerians were the people of the old age of Atlantis spanning the world, the antiquity before the antiquity, with people living throughout the lower latitudes of Earth as it is the case during all ice ages, since the upper latitudes are covered by ice. While Mu is the warm age further in the past, before Atlantis and the ice age. Mu is the warm age of Earth, or the stone age as it is named currently by science, since there are only stones and rocks left from the stone age to confirm it, with the humans of those very old times called 'hu-Mu-ans,' or humans, inhabiting the upper latitudes or the higher lands, since the increased water levels made the old territories uninhabitable underwater. Which means that there is a continuation from one age of Earth to another, from one civilization of Earth to another, or from a warm age of Earth to an ice age and then back again to a warm age, yet since you must move around continuously throughout the millennia because your land is either covered by ice or by water, all vestiges are destroyed, and therefore the current consensual authorities can claim anything that they want about the human past. However, the people of the past ages of Earth were able to keep track of everything, and so should we.

Sumerians, along with the deities of Sumer, had managed to continue their old civilization. They inhabited slowly higher lands, they prospered and thrived, and when their land became too small for them, they had to expand globally. We are already in the age of kings, the age of regular people ruling the lands in the name of their deities. Who were these deities? What did they do, and where are they currently? It is not hard to find out, because normal, natural life, and natural living societies following a normal, natural, living spirituality are capable to develop you in all details, physically, spiritually, socially, and cognitively, to very high levels. This seems to have happened until the last age of Earth, the Age of Aries, and it certainly made the difference between the people that were more or less developed on Earth. These were the old deities, as we find them in all pantheons, since they were the same, while as we notice, even after the decay of the Age of Aries, the descendants of the old people of Earth were still highly capable spiritually, yet with a lesser responsibility and with drastic consequences, while these were the demi-deities.

The Bible calls them the Nephilim, the sons of the deity on Earth, or those who lived in heaven and descended on Earth. Genesis calls them 'the offspring of the sons of the Deity and the daughters of men,' while messiah has a similar origin. This dual origin, heavenly and Earthly, seems to define currently the entire humanity, since the human genetic background homogenizes throughout millennia.

In the Hebrew Bible it is stated that: *'now it came about, when men began to multiply on the face of the land, and daughters were born to them, that the sons of the Deity saw that the daughters of men were beautiful; and they took wives for themselves, whomever they chose. Then the LORD said, "My Spirit shall not strive with man forever, because he also is flesh; nevertheless his days shall be one hundred and twenty years." The Nephilim were on the Earth in those days, and afterward, when the sons of the Deity came in to the daughters of men, and they bore children to them. Those were the mighty men who were of old, men of renown.'*

Note the resemblance with Sumer. Anu was the deity of the sky or heaven, and his consort, Ki was Earth. Their offspring were the Anunnaki, which literally means 'Anu-nna-ki' or the union between Anu and Ki, between the deity and humans, forming the Nephilim.

Heaven is a different plane of existence, a higher reality, and if the Nephilim inhabited it, this means that they could have descended on Earth only as spirits, embodied as humans or not. While from higher sources of the Bible, we learn that one third of the higher beings of the higher worlds came to Earth and intermarried, as stated above. While throughout other books of this series "Human," I study how the higher beings of our upper realities come here, why, and what they do. Yet by having one third of the higher beings or souls of the higher world coming here on Earth, this makes Earth very popular, not exactly through human achievement, but through human decay. Souls love drugs and tyranny the most, which is controversial in the higher worlds, while also frustrating our Creator, who always destroys his worlds consequently.

We also notice this world resembling to the current social media or to current videogame worlds, yet significantly more developed, which means that the higher beings or souls calling themselves deities or not cannot come in this world in person, but only as spirits or souls, coming into living human beings as souls. The Creator himself disallowed it, yet they still incarnated or came into the humans of Earth without permission, they compromised the human beings themselves in this manner according to our Creator, and consequently, our Creator ended the entire human world, or will end the entire human world soon if they refer exactly to this world.

This entire story clarifies, uncovering who the humans and the souls are, along with their entire soap opera of intermarriage and further incarnation here on Earth, since they did and still do everything as divine beings here on Earth, while they are not too divine in the higher world, but ordinary living beings, also human.

There were two types of humans here on Earth in the old times, the human avatars that you could always incarnate to come in this world as a soul, and the free humans, but you could not come into them as a soul. Earth became gradually popular in the higher world, many souls wanted to come here but there was no room left since all avatars were taken, and they only watched the other souls as they lived life here on Earth, exactly as you watch others play videogames in the media, but you are only a watcher. The watchers themselves made an agreement to incarnate or to come into the free humans of Earth without the permission of our Creator, they did so yet they compromised everything, and consequently, our Creator destroyed that world or will destroy this world very soon, if they refer exactly to this world and not to the ones before, since our Creator had already destroyed this world five times in the past and distant past, for similar reasons. While in general, our Creator tends to destroy his worlds if they decays into dark ages, for lack of fulfillment and compatibility with his intentions.

It is similar in the videogame "The Sims 4," where you can interact as you please with the sims of the other people playing them, or you can interact with townies sims, which are the free sims played by the computer itself, lacking a player or lacking a soul. Townies are rather dull and very predictable, lacking ingenuity continuously, since they do not have a human player behind them. However, you can always keep these townies sims around you throughout the videogame, for various reasons, while you can even marry them, and you can have children with them, since you can do as you wish.

Currently, Earth is even more popular in the higher worlds, making the souls to come here by the billions, with not too many people available to choose from, and you must stand in line or you can incarnate into pets instead, just to be with your soul mates in your favorite world. Highly consistent and very seductive the human reality is, because as long as they have drugs, cruelty, terror, and tyranny here on Earth, the souls will always come by the billions in order to trash this world some more, depicting the lack of development in the higher worlds. Yet since our Creator also made many of the higher worlds above, he ends everything for similar unfulfilling reasons. Yet since the even higher beings are also ignorant and undeveloped, it never ends.

Yet this world had known better times, the golden human ages. We learn from the Bibles of their mass exodus, as all souls had left in mass at the end of the last age of Earth, at the end of the age of Aries, after a very successful continuous living on Earth enjoying art, science, interconnectivity, literature, learning, and spiritual abilities. Yet they left suddenly, as though they had all been caught doing something illegal here, as though they lived their life on Earth covertly and at the intelligent human level, hiding from the Consensual Matrix. With the Consensual Matrix dispersing them suddenly, while leaving behind some souls to disconnect the people of Earth and to keep them astray and underdeveloped until the last day, the day of judgment, if this means anything.

By studying this world closely, we notice it of a very high existential resolution, allowing higher development in all details, as though our creator had development and intelligent achievement in mind when he created this world. Did he create this entire world against the Consensual Matrix?

What exactly is this last day of judgment? Why would anyone ever interfere with the way you play your videogame "The Sims 4," online? Why should they ever keep you astray, diverted, distracted, and underdeveloped? What exactly is wrong with art, music, plays, painting, dancing, spirituality, reasoning, creativity, and intelligent human social interconnectivity? What exactly is this human reality?

The Qumran is one of the old manuscripts among the Dead Sea Scrolls, not included in the Bible, or probably included and later removed. The Qumran states that the offspring of Seth rebelled from the Deity and mingled with the daughters of Cain, while the Deity has condemned the Nephilim for their rebellion.

Many books, including the Aramaic and Greek Book of Enoch, refer to the Nephilim as the offspring of angels, and this is a valuable piece of information. The Book of Enoch had been removed from the Bible, or it had never made it in the Bible. Enoch lived before the deluge, and was a great grandfather of Noah, while the Deity gave Enoch the power to see clearly. This is certainly a higher ability, but in this context, we notice how it was relatively normal to have higher spiritual abilities in those very old times, remnant higher abilities of the ancestors who lived throughout the past golden ages of Earth. Enoch saw clearly, and therefore he wrote the entire Book of Enoch. Enoch states in his Book that angelic beings mated with humans, the fallen angels, mostly Watchers. Samyaza, an angel of high rank, is described as leading a rebel sect of angels in a descent to Earth to come into the human beings: *'And it came to pass when the children of men had multiplied that in those days were born unto them beautiful and comely daughters. The angels, the children of the heaven, saw and lusted after them, and said to one another: 'Come, let us choose us wives from among the children of men and beget us children.' And Semjaza, who was their leader, said unto them: 'I fear ye will not agree to do this deed, and I alone shall have to pay the penalty of a great sin.' And they answered him and said: 'Let us all swear an oath, and all bind ourselves by mutual imprecations not to abandon this plan but to do this thing.' Then swore they together and bound themselves by mutual imprecations upon it. They were in all two hundred; who descended in the days of Jared on the summit of Mount Hermon, and they called it Mount Hermon, because they had sworn and bound themselves by mutual imprecations upon it.'*

Note the resemblance with the text from the Hebrew Bible. It is the same higher story, yet it is different, since the perspective had changed. Now it is the perspective of these higher beings. Enoch saw clearly, which means that he saw everything through the awareness of higher beings, directly through their own perception, reasoning, and therefore perspective. There are also two distinct entities referred to, or implicitly referred to as the Deity, one from each text. It is very important in understanding who these beings are, and who they claim to be. Up to now, the story seems clear: a group of higher beings had deserted their supreme ruler and descended on Earth where they lived not as mortals the way all sources claim, but they lived as deities, being venerated in real shrines. Not only this, but they claimed to be the Deity himself. However, a higher deity, their own supreme ruler, who could be the Deity from the Bible, sent an army of angels to destroy them or to take them back.

There are remnants of ancient nuclear wars just southeast of Sumer. There are also records in the Bible and in the Indian mythology and religion of wars fought by heavenly deities here on Earth, while these are so consistent, everywhere, not only in these records, but also in people's testimonies, as though this is a very important higher incident involving Earth. Yet both texts presented above had been altered significantly throughout the multitude of translations along the centuries and millennia, even along the last decades, to become what they currently state about the daughters of men.

After vanquishing the wrongdoers, the Deity still kept ten percent of these fallen angels on Earth to lead humanity until Judgment Day, and the Deity did so on humanity's behalf. The fallen angels had altered the ancient people of Earth along with their civilization in every manner, according to their interests.

I once saw a black bear at a zoo. It used to be the pet of a family who had found him in the wilderness. They took him home and they pulled out his teeth and claws, to render him harmless. The authorities intervened and recovered the bear, yet they could not return him to his environment, since he lacked the means of subsistence, having no claws and no teeth. What could the authorities do but keep the bear temporarily at the local zoo, before they found a solution? While there had never been a solution.

What could the Deity do with altered human beings, now vulnerable, lacking the means to access higher knowledge, to use higher abilities, and to ascend to higher planes of existence? The supreme leader could be the same with the Deity of Christianity, yet it could be otherwise, since all powerful beings seem to call themselves the Deity once they come in contact with Earth.

According to the texts above, the fallen angels who begat the Nephilim were cast into Tartars, a place of 'total darkness,' with ten percent of the disembodied spirits of the Nephilim to remain after the flood as demons, and lead the human race off track, until the final Judgment. this is what I always encounter throughout my studies, humanity astray and diverted deliberately, with all domains of our civilization doing the opposite: economy wastes resources, medicine kills people, food, water and air poison people, education causes lack of reasoning and therefore ignorance, finance impoverishes and bankrupts people, justice condemns and imprisons the innocent, while many false religions cause souls to perish, while this is what it means to be astray. Is the Consensual Matrix causing our Creator to end all his worlds?

Their supreme leader is not the Deity, since he is not the one performing the final judgment for the human beings during the judgment day. Furthermore, he seems to have the intention of sabotaging our entire civilization. For what purpose? For their survival and subsistence, since higher abilities are always a menace, and the young humanity had to be kept at bay, with the teeth and claws removed, because many higher civilizations harm humanity continuously in order to keep it weak, undeveloped, vulnerable, and therefore exploitable.

Who exactly are these people? At a closer study, they seem to be a military or police higher authority protecting a primitive or recovering humanity, while interacting with the people of Earth in an illegal manner by their own laws, yet they interact in a casual manner according to their natural needs. This is the difference between the real and the consensual, because throughout the consensual, you must always fulfill something else, something consensual, in place of your actual human needs.

These higher beings have to fulfill two contradictory types of needs and meanings, natural and consensual. They were in the Consensual Matrix and in a significantly higher civilization compared to humanity of those times, since while they fulfilled their natural needs and meanings for Life, they also had to fulfill their artificial needs or duties for the Consensual Matrix back home. Therefore, it is not the Deity that they refer to, but it is the supreme ruler from within the Consensual Matrix that they obey through their servitude and consensual needs. It is the Consensual Matrix, as it interacted back then with a primitive humanity, contorting it and engulfing it wholly, to have only dark ages from then on.

These are the religious characters of the common religions, the good ones and the bad ones, the angels and demons, the good and the evil, since these have many names. This is why they help and protect humanity from the Consensual Matrix, and this is why the Consensual Matrix forces humanity to hate these specific religious characters, because they went against the rules and duties of the Consensual Matrix.

In the Aramaic culture and mythology, the term 'niyphelah' refers to the Constellation of Orion, while the term 'Nephilim' refers to the offspring of Orion. I state this reference here because the descendants of the Nephilim from the Upper Egypt believed that their home was in the Duat, a place in the sky that include the star Sirius and the constellation Orion. Sirius and Orion are the representations of Isis and Osiris, yet all spirits and deities from the Egyptian pantheon lived in Duat, while all living beings, including the pharaohs, pass through the Duat and are judged when they die. We know that the pharaohs of the Upper Egypt always communicated with the powerful beings of the Duat. These beings seem to have controlled all decisions that the pharaohs made, while the pharaohs themselves were considered to rule in their name. Coincidentally, the Earth was in the age of kings then, with humans ruling in the name of the deities.

Long ago, there was a court hearing for a dispute between human beings and another civilization, separate from humanity, probably one of our current underground reptilian civilizations. These deities from the Duat had judged us both, and they agreed in favor of humanity to have the right to stay and inhabit the surface of Earth, to make use as pleased of all resources, and to rule as pleased above all animals. When the other civilization stated that Mother Earth was not allowed anymore to give birth to civilizations including human civilizations, since she had enough according to their consensual agreement, these beings from Duat replied that the current human race is only partially born of Earth, since humans are only partially native, while they are mostly descendants of people from the Duat itself. This event is also stated in the Bible, when the Deity banished the serpent to crawl underground for the rest of its existence, while he allowed the humankind to inhabit the surface of Earth and rule above all animals as pleased.

Furthermore, humans are three beings as one, mind, body, and soul, with one of them to be holly, the soul, which is what many religions call 'the holy ghost.' This holly being, regardless if it has its origin on Earth, in other constellations, in other galaxies, or in other planes of existence, is highly respected everywhere, in many spheres, and this might relate to the Nephilim, the offspring of the Deity himself on Earth. Note that, according to all these texts, humans are the Nephilim, being both corporal and spiritual beings. If you ever hear in the news of very old people being referred to as receptacles, they are considered only as holders of souls.

Another record reaching us from the very old times of Sumer is the Epic of Gilgamesh. The name of Gilgamesh from the epic story used to be Bilgamesh. Currently, the scientific consensus associates Gilgamesh with the fifth king of Uruk, who was called Gilgamesh, and ruled for 126 years at around 2500 B.C. This epic had been transmitted orally throughout the Sumerian history, and then it was recorded on clay tablets later on when writing became available. The epic is timed after the deluge. Gilgamesh was three fourths deity and one fourth human. The epic follows Gilgamesh, king of Uruk, in his quest to explore and defend his city against external threats, in the company of his friend Enkidu, a mighty, powerful giant probably not human in nature, but of a different intelligent species, always resembling closely a sasquatch.

Study Enkidu closely, to find him resembling a Bigfoot, sasquatch, or Chiubaka from Star Wars, since he was big, muscular, and furry. Bigfoot, sasquatch, and yeti are real, they are of other intelligent species inhabiting Earth alongside us, now extinct, or almost extinct. If this is true, if the intraterrestrials claim that human beings are not of this planet, then at least bigfoot is native, or they might have also come from other worlds, the way reptilians also claim. Yet it is more possible that the reptilians of Earth are ancestors of humanity, mostly since they have similar souls at times, while they are also compatible.

Now we have the chance to find out the truth about our

origins, with the Epic of Gilgamesh being the oldest record still around, presenting how our world and civilization were before, during the deluge and immediately after. We learn of at least four separate intelligent beings inhabiting Earth in those times: the deities as the deity Ishtar herself from the epic, the demi-deities, the uncivilized humans who lived everywhere in the wilderness, and the sasquatch as Enkidu himself. Yet Enkidu was the only one of his kind, throughout the entire epic, as though his entire kind was already extinct.

As you study the Epic of Gilgamesh closely, you notice how Enkidu was not exactly primitive, even though he had fur, but he was capable of intelligent reasoning. He follows good moral values and he is always ready to fight in the name of the goodness, the way all the good characters fight in all stories, always winning in the end.

As you study the ancient maps dating from the Ancient Egypt, you notice the lands of Egypt, Greece, and Canaan clearly marked on the map, along with all the other civilized nations of that time. While most of Europe, Africa, and Asia are blank and unclaimed, with only little pictures of a naked man and woman drawn occasionally. Those lands were not civilized, were not claimed, and were inhabited only by wild animals and by uncivilized human beings living in wilderness everywhere, yet they were also human.

It was the same in the Sumer of King Gilgamesh, since Sumerians lived in cities as Uruk, with uncivilized human beings living in wilderness everywhere else. Uruk was the first city ever to exceed 50000 inhabitants, being fortified for the first time also by King Gilgamesh. Between these cities, you could still find uncivilized human beings. You could even see them as they always came from the woods to drink water from the ditch around the fortifications of the entire city, fortifications used to defend the civilized against the uncivilized.

While studying the Epic of Gilgamesh, we notice how it changes perspectives suddenly, because many times, it is written from the point of view of the uncivilized human beings, while other times, it is written from the point of view of the civilized inhabitants of the city, or from the point of view of King Gilgamesh himself. This is always the case with very old records, because they are always altered throughout the dark ages, in every consensual manner. While it is rather unlikely that King Gilgamesh was the actual hero of the epic, Gilgamesh himself, since the entire epic precedes him.

As it is the case when the mythical hero Gilgamesh returns from his epic journey of finding the plant of immortality and he finds the deities living in a fortified city. Coincidentally, the fifth king of Uruk, also called Gilgamesh, was the first king to fortify a city, since they had just started to fortify their cities, in order to protect themselves from other cities and from the uncivilized human beings living in the wilderness.

Notice who the deities are, since while the civilized human beings referred to themselves as deities, they referred to the uncivilized humans as animals. Furthermore, they referred to themselves as being civilized, while they were only consensually civilized, but not actually intelligently, socially, and spiritually developed, since even the uncivilized humans of the wilderness were more developed socially and spiritually than they were, since this is why their children with the 'uncivilized' humans were even more capable than they were. There were many civilized humans leaving the cities to go live among the uncivilized ones, to sleep with the 'animals,' as they were called. It is similar with the two religious texts studied above, because humans will always do anything in order to fulfill their human needs, and therefore they will always go to live anywhere else with everybody else only to fulfill their needs, since Life herself always demands and expects human diversity, human fulfillment, and human success.

This still happens, since you will always see wealthy tourists visiting and living in undeveloped and developing nations, while they could have remained in their own developed nations to enjoy a better lifestyle.

There are other consistent references. The Nephilim of Sumer depicted the humankind as primitive, behaving as animals, since they 'drank water from the ditch, with the other animals.' These were the two groups of humans inhabiting Earth in that very distant past, the civilized humans, members of the civilized Earth, and the uncivilized ones, yet still normal human beings, living in the wilderness. The civilized group of people considered the uncivilized group to be animals, not people, yet they were always genetically compatible, since intermarriages were possible, yet highly forbidden at times. Later on, the uncivilized people became quite attractive, and intermarriages became common.

If you travel to the Amazon forest and if you find undiscovered uncivilized people, you find them indistinguishable from the people of Colombia and Argentina, they are normal living human beings. It is the same everywhere, since all these uncivilized people living between the cities of the distant past resembled exactly the people of the civilized cities near them, in a coincidental manner, making everybody similarly human, civilized and uncivilized alike, but not deities. Furthermore, etymologically, in all Latin languages, civilized means people of the city, or inhabitant of the city, while the words civil, civilized, and civilization come from the word civic, city, or citadel, because these words originated exactly in these distant times, yet in South Europe, during the times of Atlantis. The Latin language itself was contemporary with Atlantis, or it was the actual language of Atlantis, the language of the antiquity before the antiquity in South Europe.

Humans are intelligent in perspective, since humans have a cortex, yet humans cannot become intelligent if they do not learn to master an accurate intelligent conceptual language, helping them communicate, learn, and think in an accurate conceptual intelligent manner, while allowing them to reason accurately and intelligently. While without intelligent reasoning, humans cannot interconnect intelligently in society and cannot form a genuine intelligent human society along with an intelligent human civilization, necessary to maintain entire successful fulfilling golden human ages, instead of the dark consensual tyrannical ages that you know well.

Therefore, it does not matter if you have ancestry in the civilized people of Sumer and Southern Europe or in the uncivilized people living in the wilderness between the very old cities, because if you are not intelligent, you are not of a golden human age, but only of a dark age, either tyrannical or in servitude. Yet since the current tyrants claim ancestry among the first civilized people of Earth for exploitive purposes, they alter the human meaning in life and in the world, and in this manner, the dark ages never end.

Abstract intelligent conceptual language and abstract conceptual intelligent reasoning develop only through education, during childhood. More precisely, if you are born among wild animals in the jungle and you never learn to talk, you will never learn to think in an intelligent abstract conceptual manner, and you will live your life as any wild animal, regardless if you look human. It is a stereotype to assume that if you are born human, you are become intelligent by default, since you take for granted all the education that you receive at home and at school throughout childhood. Because intelligence is an achieved cognitive ability for humans, and you can never develop it if you live in an unintelligent environment.

This might seem irrelevant in an undeveloped consensual world, yet once humanity encounters a major calamity, and once civilization is not possible anymore in the current mode of civilization, with schools, mass media, and modern medicine included, but only in a very low level critical mode of civilization, based on pure instinct and on the law of the jungle, lacking basic schooling and basic justice, then people downgrade consequently, the human population decreases drastically, the human niche itself narrows, since the undeveloped humans cannot reach it anymore, they become prey to many wild animals who become more numerous by taking over the human niche, and in a matter of centuries, the current civilized people end up living similarly to the uncivilized humans from our ancient epic.

Because intelligent reasoning is not an innate cognitive

ability for humans, but only achieved, yet you need more than your strong determination to achieve it, since you must be born and you must live your life in an entire intelligent human environment teaching it to you, which is the third level intelligent human environment, possible only throughout entire intelligent golden human ages of Earth. While after you grow up and after you develop intelligently, you must teach it to others, in order to maintain the entire intelligent human environment on Earth stable, throughout all intelligent golden human ages.

While as you study the human history closely, you notice how it had never been the case, even though we still find a multitude of traces pointing to very distant intelligent golden human ages, which are systematically erased. Study Latin closely, to see how it offers the entire linguistic matrix of an artistic intelligent human language, with many intelligent human terms already included. It is similar with the entire trigonometry of the last golden human age, which includes not only accurate numbers, but also an entire developed number system of twelve or sixty digits. This is why there are twelve hours and sixty minutes and seconds, and this is why circles have 360 degrees, because everything is based on a twelve or sixty digit system of numbers, not on the ten digit system of numbers that you know well.

During unfavorable environments, as after major cataclysms, people have to fulfill mostly lower level physiological needs, since if you do not fulfill these, you die of starvation or of lack of proper shelter, neglecting your third level intelligent human needs for learning and development. This is how, in only several generations, you end up living your life at the animal level, neglecting entirely your intelligent human needs and meanings for human equality, art appreciation, and intelligent development. Without these, you cannot live your life at the intelligent human level, while if you do not live your life at the intelligent human level, through the basic civilized means that you are most familiar currently, you cannot teach your children to develop intelligently, they will never be able to live their life at the intelligent human level, neither will the rest of humanity.

However, you always have your intelligent human needs and meanings in your cognition since these are innate to humans, as these always force you to develop yourself at the intelligent human level, and through you and your intelligent human needs and meanings, they force you to develop the entire world, with the intelligent human environment and all golden human ages included. While as you study closely the human reality both real and consensual, you notice how the Consensual Matrix destroys continuously the intelligent human environment on Earth mostly through the current consensual Brotherhood, rendering and maintaining everybody undeveloped deliberately, Masses, Brotherhood and Elite alike. Because you cannot fit in the Consensual Matrix as an intelligent living human being, and you must always be ignorant, consensual, and obedient.

Now we understand why the world is still divided consensually into the same 'civilized' and 'uncivilized' humans, as we understand who the Arians consider themselves to be out of these two groups.

Yet the Arians are not the only ones claiming an exclusive civilized status on Earth, since the invisible kingdom does the same. While the Arians used to consider the rest of the world uncivilized, the invisible kingdom still consider the rest of the world animals or even abominations. The invisible kingdom conquered the Arians only one century ago, using the two world wars, removing and replacing them in the upper society to do just the same, discriminate and exploit the world. Only that the invisible kingdom also eradicates the most developed genetic lines while leaving behind only the undeveloped, while in this manner, the human race will never develop and will never survive, perishing altogether, with all invisible kingdom, Russians, Arians, and the Chinese included.

Yet even with all developed genetic lines still around, and even with humanity continuously in the intelligent golden human ages, you are never sure that you survive the strict environment of the humankind, since Life herself is always unpredictable and always unforgiving if you only blink as a species in an old, undeveloped, and unsuccessful manner. Yet with all developed genetic lines around and with humanity developed at the intelligent level continuously, you still have a chance, because humans are also psychic and spiritual at the intelligent human level, and it certainly helps.

This is how Enkidu dies, or the entire sasquatch species, yet not in a natural manner as depicted above, but in a consensual manner, because the deities themselves decided that two intelligent species on Earth can become very powerful, mostly while cooperating together since this is how they defeated the Bull of Haven, so they decided consensually to exterminate the sasquatch, and Enkidu died, while leaving only the humankind on Earth. The same powerful people who decided that the reptilians or the snake must crawl underground, while humans must roam the surface of Earth, decided now that the sasquatch must go extinct at once, while leaving only the humankind on Earth.

After Enkidu dies, Gilgamesh goes in a quest to find the plant of immortality, and he meets Utnapishtim, a legendary character who lived before the deluge, and to whom the Deity granted immortality.

The Epic of Gilgamesh creates successfully a genuine model of the world, which satisfies people's need to know about themselves and about the world they inhabit. In the first part of the epic, Enkidu clearly sees and describes the underworld, the netherworld, the world of the dead, or the Duat, while later on in the epic, Utnapishtim tells teachings and stories from before the deluge.

Throughout the epic, we find a constant depiction of the deities of the Sumerian pantheon making a living among mortals, among human beings. The mother of Gilgamesh is a genuine deity, and she lives in Uruk, while other deities live in a fortified city not too far away. In the epic, Gilgamesh and Enkidu fight the Bull of Heaven, sent by the deity Ishtar to kill them in revenge for Gilgamesh rejecting her in bed. The Bull of Heaven is the symbol of the Age of Taurus, when people prayed to a Bull, as to the Minotaur of the ancient Greek or the bull from the story of Moses.

The Sumerians had invented the zodiac, at least partially. Since Gilgamesh and Enkidu had defeated the Bull, this means that the Epic of Gilgamesh is timed at the change of those ages, before or around five thousand years ago, or before 2300 B.C. The Deity Ishtar and the rest of the Deities considered the slaying of the Bull of Heaven to be a real threat, they were already living in a fortified city fearing the uncivilized humans, and if the uncivilized humans had turned against their symbol of veneration, the Bull, soon they were going to turn against their deities, the ones living in the fortified city. The civilized humans contacted their deity Anu for divine judgment, and Anu punished the two heroes, condemning one of them to death, probably by choice. Enkidu died in this manner, and this determined Gilgamesh to begin his search for the plant of eternal life and divine knowledge.

That the decision made by Anu is not a simple punishment, because among all people, only Gilgamesh and Enkidu, together, dared to stand against the deities, with Anu deciding to eradicate one of these species, to weaken the servants and the slaves.

Currently, you can find sasquatch if you visit the most remote places of Earth, yet if they find you first, you might be in trouble, since it seems that they know that they are exterminated.

What happened with the deities of Sumer? Were they called back home by Anu? Records give all answers to this question. Some say that the deity removed them by force in a spectacular war, with humans also participating bravely in it or only watching the war as it took place in the sky, while other records say that humans had turned against their deities and destroyed them. There are records coming from beyond Earth, telling of a group of young, powerful beings maintaining order throughout the galaxy, called the Renegades, and these came and removed the deities. The deities of Sumer could have been anything, from advanced spirit beings to advanced extraterrestrials, or even a remnant of the global civilization from before the deluge managing to escape the eradication. Some say that the deities of Sumer are still in the hiding now, or they are still controlling Earth from the hiding, always waiting for an opportunity to come back in the open to lead humanity as deities again. Whatever happened with the deities of Sumer, when Alexander the Great entered Babylon in 331 B.C. on the verge of his tyrannical conquest of the world, he found the last deity ever, in the palace of Babylon, dead and well preserved in a barrel of honey.

Alexander the Great himself claimed to be a demi-deity. Before he went out to conquer the world, he visited the Oracle of Delphi, where the visionaries confirmed that he was a demi-deity indeed, since his father had been a genuine deity. This is why Alexander went to war, to reconquer the lands of his divine ancestors, to reclaim and take back what was his, the entire world, as all tyrants would always do. He was successful, yet he was too late to save his people, since they were already dead or gone, because the uncivilized people of the ancient times had become civilized, and got them. The fight for world domination had not ended then, and it still goes on, with one group winning more than another, and the fight is tight.

We know that ancient people were obsessed with observing the sky. They knew all stars and planets, and they always looked for intruding space objects. Planets look as stars, yet they change their position slowly among stars. We can spot artificial satellites currently, moving slowly across the sky. Ancient aliens could have placed their spaceships or space barges on the geostationary orbit of Earth, in which case, they could have looked just as stars. While if they were placed on lower or higher orbits, they would have crossed the sky slowly, the way artificial satellites do currently. We know that stars, planets, and artificial satellites have about the same magnitude and luminosity, with some being brighter and some dimmer. Ancient people called planets 'flying stars', or 'wandering stars', and they could have called eventual artificial satellites of Earth, 'planets of the crossing', since they crossed the sky repeatedly, while orbiting around Earth.

Yet all satellites of Earth do not have to be artificial in nature, since rocks of all forms and sizes orbit Earth right now, and you can see them clearly as they cross the sky at night. These are the natural satellites of Earth, called moons. They do not have to be very large in order to see them, since as they are illuminated by the Sun, you can see clearly even objects of the size of cars, people, and basketballs. You watch them crossing slowly the sky, and then if they disappear gradually, it means that they pass in the shadow of Earth. In this manner, you can estimate where the Sun is, in order to be able to estimate where the actual shadow of Earth is for the specific orbits of Earth, lower or higher, and you can even figure out where these objects orbit Earth. You can even see their form, since you can see their shadow casted on the Moon, as you can also see shadows of airplanes casted on the Moon. Do a search on the Internet for all these pictures and videos, to see how most of the satellites that you see at night are natural, not artificial as they should be. Because with tens of thousands of artificial satellites that Earth currently has, where exactly are they, if they are not in the sky, in the orbit of Earth?

The artificial satellites of Earth were supposed to fill up the sky, crisscrossing it at all speeds and in all directions, while there is not much up there to see. Does space even exist beyond the lower Earth orbit? Does space even exist in the videogame "Horizon Zero Dawn?" No, because you do not have to fill in all details of a created reality, but only everything that is needed, with the rest illusionary. There is a Moon in the sky of the Horizon videogame world, yet it is not a real space object of the videogame, but only an image in the sky. Coincidentally, the Genesis states that the Earth is plane, fixed on the firmament, with all images of the sky going around it once a day, which is possible only if all space objects that you see in the sky are only images and lights, but not actual space objects, since you never need them as material space objects throughout your Earthly life.

While reading the Bible, you notice that the term Nephilim refers more to spirits from other realms than to aliens living in old space barges revolving around the Sun. When the Bible says: 'let us go down' or 'and they went into them,' it refers to normal higher beings or souls performing normal incarnation on Earth, while this happens to everybody, because Earth is so popular in the higher worlds, that all humans have a soul, while most of the human souls also have souls.

Note that the 'sons of the Deity' became compatible with Sumerians during those times, or starting with those times. Once the civilized and uncivilized groups of people intermarried, they saw that the descendants of men were 'fair,' which means proper, and they went into them, as souls, not as a sexual act the way all texts describe. This means that they could not incarnate before, which further implies that, somehow, the human mind and body had either developed naturally, or they had been artificially modified to form a proper matrix for the more advanced souls coming from a more advanced realm, probably the realm of our Creator and further above. This is why the sons of the deities on Earth, or the demi-deities became men of renown.

The Anunnaki were also genetically compatible with the native humans of Mesopotamia, and this implies a common origin. More forbidden knowledge surfaces today referring to intraterrestrial beings of all forms of life, also compatible with humans. Furthermore, these intraterrestrial races, reptilian or human, call themselves Solarians, and they claim to have been here long before humans. Yet since we are compatible, it implies that we have a common origin throughout the Solar System, and therefore we have always been here.

Space barges in the asteroid belt can sustain intelligent life for significant longer than civilizations last on Earth, but probably not as long as subterranean bases can sustain life inside Earth and inside other planets.

Aliens can come to Earth from everywhere, since Life is omnipresent in the universe. It is another stereotype to assume that intelligent life is possible only at the surface of specific very rare Earthlike planets. It is a privilege and a luxury, to live your life entirely on the surface of an Earthlike planet, to walk and sleep right under the stars. Furthermore, it is a privilege to keep all the problems of the universe out of your awareness and assume that nothing else exists out there, nothing to harm you and nothing to exploit you in any way. However, intelligent life is present everywhere, on the surface of habitable and uninhabitable planets and moons, inside moons and planets, which is most common, and on artificial space objects, anything from very large artificial structures comparable to moons and planets, to regular spaceships, space stations, or space barges.

As another stereotype, aliens come to Earth to seek resources of any kind, things that have more or less value to us, including simple salt, water, or gold. This is a stereotype formed throughout the years by our entire collection of science fiction books and movies, because materials and resources are more abundant and easier to harvest out in space than on Earth. Simple machines do a better job harvesting them, since while using humans to do the work, it is as making elephants work in the fields and forests. You can always use elephants to move logs, but you still have to supervise and control them directly, one on one, while robots work independently. Humans are also very unpredictable, unlike many animals, and very treacherous, since humans will kill anyone unexpectedly for any reason, to the point where you run out of masters very fast, if you keep exploiting humans. Only humans can exploit humans currently, exactly as it happens at the lodge with all handlers, yet even so, it takes an entire Consensual Matrix to keep humans under control and in exploitation, since otherwise, humans return to the fulfillment of their own intelligent needs and meanings, while making entire golden human ages possible once again.

Throughout time, humans had killed aliens along with everything resembling to aliens or deities, for freedom, justice, virtue, revenge, and more, because it is said that eating the flesh of the deity gives you supernatural powers, or it even makes you immortal. Humans actually eat aliens when they have the chance, even the little grays, and they taste like chicken mixed with copper. More precisely, this is what the uncivilized humans did, they ate civilized humans from the times of the old Sumerians, and it was probably very common. You can also be eaten currently, if you venture in the unexplored Amazon and Oceania.

The current Elite consider themselves deities on Earth, even if not too many generations ago, they lived among the Brotherhood and among the Masses. As a sad and ironic example, one century ago, a Rothschild member and his guides were exploring an uncharted territory in the jungles of Asia, and encountered an isolated primitive tribe. They were eaten only hours later, not because they had stumbled upon a tribe of hungry cannibals, but because they were white, clean, an well groomed, the primitives had never seen anything alike and thought that they were deities, so they ate them while hoping to gain their spirits, life force, and supernatural powers. The Family must have certainly revenged their deceased member, while you can still travel there yourself to learn all the details, yet the Family had already exterminated that entire tribe.

Why do we have aliens coming to Earth, repeatedly? Firstly, our intraterrestrial cousins living beneath have colonies inside several other moons and planets, and they constantly visit and receive visitors, while many of those spaceships are associated to them. They might also be associated with the human Elite, now that the Elite have improved their status in the wider world. Aliens can visit the Elite for various reasons, but if they come for the Masses, they are probably interested in the same thing as ever, human genes, human higher abilities, and human souls. Since this is what aliens mine on Earth, neither gold, nor humans. They also mine everything associated to human souls, past lives, spiritual values, possessions of past lives, life energy, human identities, supernatural powers, and even living bodies without souls called biological units, to be used at will. They can also be interested in vivid memories along with vivid feelings associated with full experiences, along with anything else that they can get away with. There are certainly not gold and diamonds first on their list, yet aliens can carry these for trade, if they ever feel as trading and not only taking by force. While most of the time, these extraterrestrial aliens are only ultraterrestrial, which means that they come from higher realities yet in spaceships, since life is very diverse in the wider world, if this ever makes sense.

It seems that these space adventurers, these invaders do not come here for maintenance or love, but they have specific agendas, specific interests, they are here to prey on whatever seems to be most valuable in the universe, divine human souls, with many cults and religions preying on the same souls. Whether they are simple thieves raiding a few humans before they run away, an entire organized band of thieves breaking deals directly with the authorities of Earth, or extremely powerful beings planting our entire planet as a garden with humans and human souls, through their entire activity, they interfere with humans and with the human civilization directly, changing it on their behalf, while inflicting all pain, all loss, and all misery possible. This happened with the last group of extraterrestrials and ultraterrestrials harvesting humans, they influenced us for the worst, they interfered directly with our genes, society, religion, and civilization, and they still do so, while it seems that we will never recover from our misfortune. How could we? How could we recover when they tempered with our higher abilities?

Yet life goes on, century after century, millennium after millennium and age after age, adventurers come and go, some tend to us, some prey on us, and it really seems that we will never recover.

One member of an underground civilization co-inhabiting Earth with us, which had witnessed firsthand our less-fortunate part of history, states that it is time for us to recover, to wake up, to forget the past, and start living life the way we should, since this happened over five thousand years ago, and those beings are gone now, never to return. It is bout time to move on.

This chapter has taken us from the last ice age to the present day, around the Earth, inside space barges and underground basis, chasing mighty deities of renown around the world, while helping us understand better the world around, our origins, along with what makes the world turn, in an actual human reality. Let us continue now our study of the official models of the universe, since we still have three more models to study. The big bang theory is considered the most accurate, it is currently the official cosmological model, and I am describing it shortly in the next chapter, along with the other two contemporary cosmological models, the static model of the universe, and the steady state model of the universe. We are interested in understanding what had led to these cosmological models in particular, who the scientists working on them were, how they had brought their contributions to such elaborated models as the big bang, and why. Let the study begin.

7 THE BIG BANG AND THE ENTIRE CONSENSUAL MECHANISM BEHIND

Before we start our study of the modern cosmological models, note that they are not the creation of a single individual human being, as Copernicus or Bruno, going against their contemporary scientific views and ending up burning at the stake. Because during this century and during the last one, we notice thousands or tens of thousands of very capable physicists from around the world working on the big bang theory itself, or only influencing it with their work, throughout all research facilities on Earth, underground, and in space, worth trillions of dollars, in an outstanding complex work of scientific creation, out of the need to unveil more truth about this world, while creating an entire model of the universe, which is also the subject of this book.

While the trillions of dollars spent on one model of the universe alone are enough to feed humanity, clothe humanity, cure humanity of all diseases, and educate humanity endlessly. This could replenish all shortages in the world, while saving tens of thousands of people from illnesses and starvation. Everything is about money and saving lives, while the brightest minds of this world, all physicists and all mathematicians who had chosen to dedicate their life learning and researching in order to leave something behind, in order to make this world a better place, are caught continuously in this kind of useless work, to explain that the last decimals of the value of the background radiation are consistent with the latest value of the constant of inflation, which is unhuman, and they should actually feed the world instead. Because these bright minds and wonderful souls had not come in this world for this redundant work, but to work on something fruitful, on helping humanity, as their own intelligent human needs and meanings demand. Yet there is nothing fruitful to work on, since all fruitful work is banned through the Brotherhood. So this is what they do, working on the big bang theory by the millions, which is only a theory, and it never feeds the world.

Furthermore, the current science will declare the big bang theory obsolete very soon, to embrace a new theory, a steady state one, through a very remarkable discovery, in an extraordinary triumph for science. Yet will they ever give back all these trillions of dollars that they keep spending? Because we want our money back, to have some food.

Why do people work on useless research projects? Probably because they assume that these are the most advanced projects in the world, the most tedious, and therefore the most important. Because there is a stereotype to assume that the hardest, the most tedious work is also the most useful and the most important, by default, which is not the case. Because if all these physicists and mathematicians would have worked on the best sausage machine ever, they would have found it by now, it made food available to everybody even free of cost, and they would have truly made the world a better place.

The big bang model of the universe is not the only modern model of reality. There are two more models of the universe, considered obsolete now and rejected by everybody, yet we start our study with them, since they are created by the same famous physicists that you know well, before they changed their mind to embrace the current model, the big bang theory.

The first modern model of the universe is the static model.

The Human Reality

In the static model, the universe is both spatially infinite and temporally infinite, while space is neither expanding nor contracting. The universe does not have a spatial curvature, but it is flat. A static infinite universe was first proposed by Giordano Bruno. Einstein proposed a temporally infinite but spatially finite model in 1917 in his general theory of relativity, and this was before Einstein had embraced the big bang theory. His cosmological model was not stable, since all matter tended to collapse in a central point. Einstein did not reconsider the laws used in his model, but he only added a positive cosmological constant to his equations of general relativity to keep everything stable, to counteract the effects of gravity on ordinary matter. It is still impressive to see how the 'greatest physicist ever' creates a static model of the universe finite in space, and then he remained surprised that it tended to collapse in one point, due to gravitational attraction. Therefore, he added a constant to balance the equations, and it worked. Yet is this the actual human reality. No, but it is the consensual human reality, while every time you praise Einstein, you are part of it.

Only an infinite universe is stable, we can see it beforehand, yet why exactly was Einstein interested in creating a finite universe altogether? Later on, Einstein settled for the big bang model, which is also finite in space and time. It seems that the models of the universe that our greatest physicists create are not necessarily true, but only a matter of taste or duty. Not only this, but he was good friends with Edwin Hubble, the 'father' of the big bang theory. Do a research on them, and you see that the great Families sponsoring them are high up in the Upper Brotherhood now, and high up in the invisible kingdom. It is their agenda that you always fulfill, since you cherish them throughout schools, along with your children and grandchildren, while an entire invisible kingdom has a big laugh, counting money and making profit by the trillions of dollars.

What profit and what money? Only the Masses think of money as being lost during wars and recessions by the trillions, or being spent by the trillions throughout major scientific projects including space exploration, while money is never lost, but it only changes pockets and bank accounts. Where does everything go? It is not hard to trace money, to find it shifting accounts fast throughout the invisible kingdom. While the dictators of the East have learned to do the same, transferring money in large amounts from their own nations to their private pockets through similar wars, recessions, research projects, and space exploration, while you toil your whole life throughout factories, supermarkets, underground mines, restaurants, and construction sites for decades, then when you finally retire, still dizzy of the entire experience, you wonder what went on and what it was all about. Yet if you had a soul with you this entire time, with both of you clueless of each other, now this is the actual model of the human reality, since it even makes sense.

The second modern model of the universe is the steady state model. In steady state views, new matter is continuously created as the universe expands. The steady state theory states that, although the universe expands, it does not change its appearance over time, with this entire extra matter being created continuously, and therefore it has no beginning and no end.

The third modern model of the universe is the famous big bang theory. The physicist associated the most with the big bang theory is Edwin Hubble, yet there are other physicists contributing to this cosmological model, in larger proportions, even before he did, while Hubble seems to take all the credit. Since it was the astrophysicist and Roman Catholic priest Georges Lemaître to show that the universe appears to be not static, but expanding.

Edwin Hubble used the data recorded by the astronomer Vesto Slipher to confirm a relationship between redshift and distance to galaxies and stars, which is the basis for the expansion idea introduced by Lemaître, creating what is called Hubble's law and metric expansion of space. Observations of distant galaxies and quasars show that they are redshifted, or that the light emitted from them is shifted to longer wavelengths. Hubble took a frequency spectrum of these objects and matched the spectroscopic pattern of emission or absorption lines corresponding to atoms of the chemical elements interacting with light. These redshifts are not locally or occasionally, but they are uniformly isotropic, distributed evenly among the observed objects in all directions.

This redshift was the phenomenon leading to the idea that all distant objects in the universe move away from an observer, since it was similar with the lowering of the sound frequency that we hear when cars speed away, the Doppler Effect. The scientists of those times had considered the invariance of the speed of light to be similar to the invariance of the speed of sound in air, and that was exactly what was creating the redshift, the distant galaxies speeding away at great speed, just as speeding cars, while this is only a common scientific stereotype.

Notice the great difference between the Doppler Effect and the redshift of distant galaxies, because in the Doppler Effect, cars sound at higher frequency when they approach and then they sound at a lower frequency when they depart, while in the redshift, everything is shifted towards red and therefore seems to speed away, as though Earth itself is a privileged place in the universe with everything else receding away, which should already seem suspicious. Yet if the scientific consensus accepts this specific big bang theory, then it must be a theory. It is still suspicious, for an entire world to accept this theory as being a theory. Furthermore, all scientists accept it as a theory to be true, which means that it is still a theory, and therefore not necessarily accurate, but only in theory true, so why does an entire world accept it as accurate? Is it only for some people to pocket money by the trillions of dollars?

This is suspicious again, because science feeds the world false information, yet when you study the manner in which it states the false information, science itself informs you that it is only a supposition, a speculation, a theory, and nothing else, nothing trustworthy, nothing evident, nothing accurate, nothing actually true, but only a theory, a speculation, and nothing else. It is similar with the entire act of law, because it is not real, but only an act, only a play, and you cannot have your money back, along with all the time that you spend in jail. Yet if the people of Earth believe such an act, theory, or speculation, then they do so at their own risk. Which means that the current science never gives you your money back for all the lies that it states, because the current science already informs you that everything is theoretical, assumptive, or speculative, never accurately true, while once you believe it in school and throughout the media, you cannot have your money back. Because science never lies, but it only makes these kind of suppositions, called theories. While for you, if you ever state that the multitude of theories of science are not true, then you are in trouble and you lose your job, even if you work in a factory, since it means that they cannot trust you anymore.

The use of the Doppler Effect in explaining the redshift of distant galaxies is just a basic stereotype, since the Doppler Effect had just been discovered at that time, and was capable to explain the difference in sound frequency that people heard everywhere there were speeding cars involved. Because automobiles had just been invented and were just appearing on the streets of all modern cities, along with police cars, fire trucks, and ambulances. Yet what exactly makes modern physicists assume that the only explanation of the redshift phenomena is the simple Doppler Effect? Why not searching for other explanations, as the difference in the curvature of spacetime, or in matter formation?

It is true that there are other phenomena supporting the big bang theory, as the cosmic microwave background radiation. This radiation was interpreted immediately as the remnant radiation from just after the big bang, yet it could be easily associated with a static universe just as well.

It seems that one century ago, the physicists had invented this magnificent explosion in the sky, the big bang, only to prove a simple phenomenon that astronomers were just starting to notice in the sky, the redshift. Yet there was significantly more to consider at that time, since the invisible kingdom was using science, the specific science that they had just hijacked, to oppose and therefore to destroy the specific religious ideology of those times throughout the Western Civilization, Christianity itself, because the big bang theory was used to confront the creationism itself.

It even succeeded, while taking down the old European aristocracy along with their entire world order, replacing it with the invisible kingdom. The Masses and the Brotherhood believe in science and not in Christianity anymore, and it worked. Now the invisible kingdom controls the Western Civilization through ideologies, along with most of the world, with the entire world believing in science now, and therefore obeying the invisible kingdom entirely and unconditionally. This is the case because science has priority in court, and no one can state that science is wrong. Therefore, if science states that for the past several million years, the human race originated exactly in the Caucasian region of Earth, and migrated from there to span the world and therefore the entire world is theirs, then everybody believes it. While only the invisible kingdom originated in Georgia from the Caucasian region, and migrated to span the Western Civilization or the entire world, and no one else.

Once you state that you originated in the invisible kingdom, then you belong to the invisible kingdom, and then you must obey the invisible kingdom. While you must accept everything that the invisible kingdom does to you and to your entire genetic line, while the invisible kingdom is allowed to do to you and to your entire genetic line as it pleases, even to exterminate you in every manner. Because every time you mark yourself Caucasian on all legal documents, you confirm that you are of the invisible kingdom, and therefore you belong to the invisible kingdom as property.

The big bang model of the universe is not an analytic model but only a theory, as the name states. Because intelligent, conceptual models are made from scratch, and they start with the main cause, using directly the natural laws of the universe.

What physicists did, they simply took one or two space phenomena, and forced scientific explanations to them in order to make them seem to be effects of the main physical cause that they were assuming to have formed the universe, the big bang. We observe the redshift in the sky, and therefore only the big bang could have made it possible, which is erroneous as a line of reasoning.

How many errors of reasoning are in the big bang theory? First, they do not observe the main cause, the big bang itself, but they only invent it. Then they take a few physical phenomena, they twist their causes, and they associate them with the big bang, artificially constructing the main timeline and the entire cosmological model. Nothing made sense then, and nothing makes sense today. This is why everybody works hard to adjust the entire big bang model in every manner, since it does not hold consistency, it never did, and it never will. Because the big bang theory is not a genuine intelligent conceptual model, it is not comprehensive in reasoning, and it is not even empiric, but only consensual, enforced through assumptions that still do not hold pertinent reasoning.

Science is full of these theories, while science even states that they are only theories, yet the world forces you to assimilate them as accurate facts. It is the same with all ideologies and jurisdictions, since all ideologies and jurisdictions state that they are made of ideological beliefs and legal beliefs, yet they force you to accept them as accurate facts and as unconditional truth. Because science is both and ideology and a jurisdiction, while making its own laws and beliefs consensually, called theories.

What should it have been done? You can never force models to work, in general, while you create them, but you should seek to find the main causes coming from every physical phenomenon truly involved. Do you observe redshift in distant galaxies? Yes. Then it is a loss of energy, which can come from any phenomena. Do not invent the main cause first, as very high speeds involved, but study why the red shift really takes place. Do not assume that galaxies speed away because cars had just appeared on the roads of your city replacing horses, their sound frequency changes whenever they pass by, and therefore billions of galaxies around must behave just the same while there is no air and therefore no sound involved, yet they must speed away from you through redshift alone. Just search for more phenomena in the sky, to find out why you encounter the redshift itself. We can even solve this entire circumstance right away. What other phenomenon?

There is another phenomenon related to the same galaxies experiencing red shift: they spin faster. Stars spin faster in the galaxies that are very far away, which should be impossible, because they must spin in a similar manner everywhere, closer and further away, since all laws of the universe are invariant. It could be only a matter of perspective, or only a law of physics badly stated or badly implemented, and therefore we must understand this error in order to be able to understand the entire world. These two phenomena, fast spinning stars and fast speeding away stars, should be connected, since everything is connected in the universe. While the big bang theory cannot connect them, but it only attempts to connect them individually with itself, through other invented phenomena that do not connect to themselves either, as the dark energy and the dark matter. Everything is an irrelevance, yet since you can never witness the big bang yourself, and since you can never observe the dark matter and the dark energy by default since this is why they call them dark, then nothing can ever be proven wrong. What is next? Dark momentum and dark frequency? Because there is no other lie left but these.

While as stated, it costs trillions of dollars for each lie, more than the entire fake moon landing. While entire genetic lines from poor nations are eradicated systematically through major shortages, which would cost less to tend to than these invented big bang, dark matter, and dark energy. Yet the people of Earth love it, as they acclaim the world's greatest scientists, along with the great Einstein himself, while the entire world cares less about poverty, misery, loss, discrimination, exploitation, and starvation.

Yet if the people of Earth choose to live life in this manner, by killing themselves systematically, while being entertained through theatrical cosmological research projects, then let the world be this way, with nations eradicating nations in every manner, more or less obscure. Yet do not expect to survive the next cataclysm, because there is very advanced real physics involved in each cataclysm, and by inventing lies for fame with each cosmological model of the real world, you miss your chance to find the accurate solution to save the world from the entire natural catastrophe. Nothing that you can invent then will save you, since you do not have the time, regardless of how deeply underground you hide. Yet with the invisible kingdom worthless, they might survive a week underground, or not.

When you observe a new space phenomenon as the redshift of distant galaxies, you do not start guessing in order to assume a main cause right away as the big bang itself, since you might be wrong. You want to relate more observed phenomena with the most basic structure of the universe and with its natural laws, relating it either with the spacetime continuum, or with the electromagnetic field itself. It is easy to see the entire model. Therefore, time itself passes faster for galaxies found further away, while physicists associate redshift with speed erroneously, not with time, while considering time and space unchanged. The more redshift they detect, they associate it with greater speeds of all stars, speeds so great, that the stars should never be able to hold orbit. Therefore, after inventing the redshift of distant galaxies and stars to be associated with speed, now when this speed is too great, they have to invent glue to keep stars in the orbit of the distant galaxies, with this dark glue being dark energy and dark matter. Because you need two of these, one for stars and one for galaxies, or this is what science states, while all laws in the universe should remain invariant in all systems of reference. All lies require more lies because they are never consistent, just as all turtles require more turtles to hold the Earth all the way down, in another cosmogonical model of the universe.

There is no redshift associated with speed as it happens with cars on Earth, since this entire theory is erroneous, while by using it, science answers a question with the true answer of another question, which is a basic error of reasoning.

In order to find the true model of this world, not only that we have to find our way on our own in the place where science had already failed, after employing a good part of the world and after spending trillions of dollars and counting, but we must find out where the redshift of the distant galaxies and stars comes from, since you can detect it even by using simple experiments of spectroscopy. The redshift itself is real, yet where is the error?

Furthermore, since we study the universe as a whole, we must look for a model of the universe and for a main cause to all the above phenomena at its base level, at the level of its spacetime continuum. If those distant galaxies are shifted towards the red, it means only one thing, not exactly that they speed up, which is impossible just by interpreting correctly the spectroscopy results, but that they slow down. Because it is a matter of perspectives, since within the spacetime continuum, what seems for us to speed up here through any experiment we undergo, it actually slows down there, very far away from us. This happens through the superimposed gravitation coming from all the mass in the universe, taking place in every point of the universe, making time to slow down in every point of the universe, due to the superimposed mass of the entire universe. Time slows down, while nothing speeds up.

More precisely, time slows down out there for each star of the universe, it slows down here where we are, and it slows down in all points of the universe, from the superposition of all mass in the universe affecting all points of the universe, yet you must consider everything in a relativistic manner, as it shows a redshift for all distant stars. Similarly, time slows down in every black hole, because of the very high mass, if black holes even exist. Yet you must use the law of relativity, which means that you must be in a relativistic circumstance first. While every time you consider the universe wholly, at universal scales, you enter relativistic circumstances, and you must apply the law of relativity, which affects the spacetime continuum itself, slowing down time.

Because when you consider very large distances, you are under relativistic circumstances, and you have to reason accordingly, by using the law of relativity. Since this is why time slows down, and you can use Lorentz's equations to compensate. However, Einstein forbids you to consider this circumstance as relativistic, while allowing only the three circumstances that he chose, as spaceships travelling close to the speed of light, the distortion of starlight in the vicinity of large space objects, and the slowdown of time in black holes.

From our own perspective, time always passes normally, yet from a faraway perspective, time slows down here where we are, giving us a redshift, only for them to notice. It is similar with all distant stars, since from their own perspective, their time passes normally, while from our own perspective here far away from them, their time slows down, while giving them a redshift. Because when you study larger parts of the universe, you involve very large distances, therefore you are in a relativistic circumstance, and therefore you have to use the law or relativity, neither classical physics, nor the theory of relativity, but only the law of relativity.

While with time slowing down very far away, all light coming from far away stars and entire galaxies is redshifted. While according to the law of relativity, by using Lorentz's equations, we notice how the farther away we search, the larger the redshift, which is what we observe in the real world.

Which means that the universe is static in space, but finite or infinite in time and space, with the law of relativity slowing down time to zero very far away, enclosing the universe for us. Which is only relative to us, while the universe is larger and older than what we can perceive. Yet the size and age of the universe are so large, that for us, it makes no difference if the universe is finite or infinite. Yet with no energy escaping the universe as an enclosed system in a normal reality, the universe can be infinite in time just as well.

Again, Earth is not a special case in the universe. It happens to them there very far away, since for them, it seems that time slows down here where we are, allowing them to notice a similar redshift here where we are. Everything is a matter of perspectives, and furthermore, you can use the law of relativity to model this entire phenomenon. While by modeling this phenomenon using the law of relativity, you are able to find a static model of the universe that does not collapse in itself as it happened with Einstein, since time slows down to zero very far away keeping it steady. This is what Einstein failed to find, a static steady universe, while he even added a constant to keep everything in place, in an ignorant manner, and he still failed.

As you notice, it takes you only a few minutes to model the universe by using the law of relativity through an intelligent human mind, while you get to use the trillions of dollars to feed the world endlessly, while employing this entire army of theorists, cosmologists, astronomers, and astrophysicists as cooks to prepare all free meals to an entire world, for a job very well done. Yet if you use all the armies of the world to do the dishes instead of sending them to fight and kill throughout wars, you manage to save the world both from wars and ignorance endlessly, not only from misery and starvation. Behold, we have just found the best sausage machine, capable to feed the world endlessly, exactly as promised. Yet this is not exactly the most efficient sausage machine, but this is the actual intelligent communal human life, in contrast with the current human disconnection that you must always obey.

Because when you hide to humans their own origins, as the origins of this world, of this reality, of life, of humanity, and of the human intelligence, you contort all knowledge about the actual human reality, altering in this manner the human meaning on your behalf, which allows you to harm, exploit, and exterminate them in every manner, even by constraining them to harm, exploit, and exterminate themselves. This was your last chance, since this was your last world, and never again.

Which is the actual model of the universe? It relates with

the attraction of all mass in the universe that we experience here, causing the time to slow down here where we are, as the law of relativity states. Yet this happens everywhere, in all points of the universe, similarly, even very far away, with the very far away galaxies. Time slows down even more further away, and therefore everything that you see there shifts towards smaller and smaller frequencies of light or electromagnetic radiation, called redshift.

We have explained redshift, while modeling the universe, as static in nature, and infinite or finite in time and space. The universe never collapses upon itself, because it is held in place through the spacetime continuum itself, which slows down time considerably at each point in the universe, as it is perceived from all points of the universe found very far away, because all these faraway points of near zero time rate hold the universe in place wherever you are. Everything is more complex, since it is only a matter of relative perspectives, yet it is enough to model the world, for now.

Because at scales as large as the entire universe, we can embark any line of causality present at universal scale, since all universal lines of causality lead to the main cause of everything, as existence, creation, reality, causality, life, and creation or birth of the entire universe. However, both the creation and the birth of the universe take place outside our universe, not within, since it is impossible to do so within. Only when you ignore the other realities of our wider world, you assume that the universe creates itself from the inside of the universe, or that it gives birth to itself from the inside of the universe, with its entire spacetime continuum and its entire existence included, which is an impossibility in itself.

This is as saying that your main character from your book wrote the entire book itself, with him and with all the other characters within, and with all real ink and all pages included, which is an aberration. Yet watch all these scientists how they define the exact time of the apparition of the universe some 13.8 billion years ago, a moment of apparition of the entire universe defined in its own time frame, with its own spacetime continuum included in the big bang itself, since this is the aberration, and shame on them for altering, cheating, and destroying this world. Since it means that all these physicists are ready to claim that they are capable to define the apparition of time itself at a specific time in the past defined by the same time frame that appears simultaneously with that specific moment in time, while using the same time frame that appears then, which is an aberration. It is the same for space, since physicists define both space and time to appear simultaneously with the entire universe during the big bang, in the same space and time themselves, which is an aberration.

If they consider that the universe was this infinitesimal point and it started exploding or expanding very fast, and if all space and time were within that initial point, then what space, what time, and what continuum were holding that specific infinitesimal universe as it was before it appeared and exploded? Could it be a turtle? Because you need an outside continuum to hold the exploding universe and its continuous expansion, while it is impossible to define that specific outside continuum from within the universe, but only from the outside. Yet even this is impossible, because higher realities hold their lower created realities not directly in their spacetime continuum, but in a specifically encoded matrix or buffer.

Furthermore, by the law of relativity, you cannot have mass densities higher than what we currently have in neutron stars, pulsars, neutrons, and protons, since matter cannot collapse on itself further. Therefore, all black holes are impossible, along with the big bang itself, because you cannot exceed a very specific mass density, just the way you cannot exceed the speed of light.

Science never researches higher and lover realities, but only this world, while science had already altered and truncated the law of relativity to the two theories of relativity, and cannot accept more, while making the entire erroneous big bang theory consensually valid. When exactly did the universe start expanding? Where exactly was it expanding in, if the space and time were within the universe itself? How exactly do you define an expansion from within, when you and the entire inner system of reference have to expand at the same rate with the entire universe, making everything seem static from within? Do you see how science makes the transition from the perspective of an upper reality holding our universe in its own higher continuum, in order to define our universe throughout its own higher existence as it appears and then as it explodes or expands? Yet then, science switches immediately to the inner system of reference of our universe, to model the apparition, explosion, and expansion from within, as it charges trillions of dollars and it keeps on charging, because this type of erroneous research never ends.

While as you already notice, the entire thing is worthless. Because if our universe expands, then it certainly expands out there, in the upper reality, as our upper reality is holding our universe, which is impossible again, because all inner worlds and realities are held in matrices or buffers, not in the continuum of their upper realities holding them. While from our own inner perspective, our universe should always be static even if it expands presumably, because the spacetime continuum itself expands or contracts with the entire universe, rendering everything always static.

Does the universe expand or contract? Everything depends on the current consensual science, since it can claim anything consensually, while the universe will always remain static as seen from within, just as you perceive it right now as you read this book. You cannot exit the universe in order to witness or even study its beginning, because from our limited perspective inside of the universe, we always see the universe as being infinite in time. While we cannot detect its beginning, because the beginning of the universe is defined outside the universe, since it took place outside, in an outer reality, which is our higher reality where the souls live. Yet even in our higher reality, our world is kept in a matrix or buffer formed by using computer technology and mind technology combined, but not directly in the spacetime continuum itself. Therefore, you cannot even witness any movement taking place outside our universe from within the universe and from the higher world, because our own spacetime continuum and our own existence cannot define anything objective taking place and existing outside our universe.

Yet the invisible kingdom let the people die in sickness, misery, wars, pandemics, and starvation, mostly when they own science along with the war industry, the food industry, and medicine, along with all commerce and transportation capable to bring food to everybody in the world. Because as we always notice, the invisible kingdom does not use science only to dumb down, exploit, harass, and exterminate the humankind, but it uses science to ridicule and oppress the humankind, as it does through every domain of society that it owns, from finance to art and education. Everything is used to make fun of the humankind with all these universes of thirteen dimensions curling up on themselves, as they exploit and exterminate the humankind.

How does everything take place? What they did with the big bang and with the steady state models of the universe, they attempted to model the moment of the big bang from within the universe, they did so for matter, energy, and spacetime continuum simultaneously, while you cannot do so, since you lose your system of reference, and this only amplifies the irrelevance. Yet they knew this, so they did not exactly model the exploding big bang, but only as it happened a fraction of a second after explosion, since in this manner, they could use our spacetime frame of reference from within the universe. While the explosion itself is not modeled yet, the big bang itself.

Because we keep our system of reference inside the universe, in the spacetime continuum, while continuum is relative in this world. Which means that time can pass slower here and not faster there, everything being a matter of perception. One problem causing this is the theory of relativity as it is defined by Einstein, because Einstein truncated it deliberately in order to apply it to whatever he observed then, those three phenomena. Yet we encounter the same relativistic phenomenon for billions of galaxies moving faster there, while it is more likely that we are moving slower here, experiencing time at a slower rate. It is more likely that time seems to pass slower here in rapport to distant galaxies, or they see us moving slower here, while we see them moving slower there. This difference in time is also noticeable now in all related phenomena, as a difference in energy, observed as a decrease in light frequency throughout all spectra, called redshift.

Do you see how they assume that all galaxies move faster there, instead of considering that we move slower here in rapport to them, through a slower time rate? Which is more likely to be the case, because it is easier to deal with one phenomenon taking place here in our perspective, and not with zillions of phenomena happening there and everywhere else simultaneously. The continuum is relative and therefore both cases are true, but it is easier to model one phenomenon happening here, than billions of phenomena happening there. Yet they had to force the big bang in this manner, and therefore they chose to study it there, far away and billions of years in the past, where it is impossible to be proven wrong, while this phenomenon does not happen here or there, but everywhere. Which means that it had never been an explosion or an expansion, but this world is relatively stationary ever since its creation, as it is the case with all the higher worlds above, the entire wider world.

It is no coincidence that Einstein worked on both the big bang theory and on the theory of relativity in a conflict of interests, while he avoided using the theory of relativity on the big bang theory deliberately. The big bang theory was never considered a relativistic circumstance, since if it was, it never made sense, compromising the theory of relativity itself. Yet he managed to form and distort both theories himself in any manner he chose, in a major conflict of interests. This favors the invisible kingdom, because it managed to confront creationism and the entire religion in this manner, overtaking it easily. While it favors the Consensual Matrix just as well, because the world is more consensual now, and therefore more compatible in tyranny and exploitation.

You cannot have a model of anything, including the most important model, the birth or creation of the entire world, with an inner system of reference. This specific model that we study here, the model of our world, should be at the base of any other model ever, and it certainly must be accurate. While with an inaccurate model as the big bang theory at the base of science, everything becomes consensual in nature, an ideology, but not real, not genuine, and not a fact. This is done deliberately, to keep humanity ignorant, undeveloped, vulnerable, and exploitable in the Consensual Matrix.

Because your other choice is to adopt the model of a living universe, a living human reality, part of a living wider world called Life, Intelligence, Interconnectivity, Mother Nature, Universal Mind, the One, or in any living manner you please, while once you adopt a living model of the human reality, you treasure and consider all details enhancing life, just as Giordano Bruno did along with everybody else throughout all living intelligent golden human ages of Earth. You fulfill your natural needs and meanings, through them you develop to the intelligent human level, and through your development, the entire world develops. While after doing so for generations and world ages, you develop yourself and the entire world so much, that you achieve to live life on Earth in perfect harmony, matching your human nature with your human meaning and your human fulfillment in an entire resonating music of the spheres alongside everybody else.

This is what the Consensual Matrix hides, along with its hierarchic Brotherhood, Elite, invisible kingdom, dictators of the East, and entire armies of scientists, educators, and university professors, media presenters, doctors, drug dealers, and entertainers. Because within the Consensual Matrix, you must have the perfect coordination of billions of individuals only to keep you distracted and on lower developmental levels, because the human nature is very capable and very menacing for all lower level entities everywhere, including the Consensual Matrix.

We should hurry up studying the big bang, because the current science is about to discard it in a big bang breaking news throughout the world, stating everything erroneous about the big bang theory, while adopting consensually another aberration instead, the steady state cosmological model of the universe. What can it ever be erroneous with the big bang besides everything seen? Everything, because all fakery is erroneous and inconsistent in all details, which are very numerous for the big bang.

We start with Hubble. When the redshift is interpreted as a Doppler shift, the recessional velocity can be calculated. For galaxies, it is possible to estimate distances via the cosmic distance ladder. When the recessional velocities are plotted against these distances, we observe the linear relationship known as Hubble's law.

$v = H_0 D$

where v is the recessional velocity of the distant object, D is the comoving distance to the object, and H_0 is Hubble's constant.

That this universal expansion was predicted from the theory of general relativity by Alexander Friedmann in 1922, and by Georges Lemaître in 1927, well before Hubble made it in 1929, yet you must have learned in school that Hubble discovered it. Even so, they simply assumed that that correlation is linear, and not otherwise, and they simply gave a value to that Hubble constant, whatever they found best. This is certainly not intelligent reasoning, but simple consensual guessing, because this formula was not supposed to be linear but relativistic. Therefore, they were not supposed to use a simple constant H, but an entire relativistic equation, made by Lorentz. Because it involves universal scales, which is a relativistic system of references, and therefore it was supposed to contain Lorentz's relativity transformation.

Yet if you use the law of relativity, then there is no big bang involved, since nothing really moves away from anything, as it is only a matter of relative perspective. Just the way the entire world moves backwards when you ride in a car and you move forward, yet you never claim that the world moves backwards, but only that the car moves forward, with you in it. Nothing moves away from us but only our time seems to slow down here for various reasons. Our universe is a static universe, and therefore there was no big bang to begin with.

Religion and spirituality believe in a creation model of the universe either natural or artificial. It is a static model, and there is never a big bang involved, but the Creator conceived it, imagined it, designed it, formed it, encoded it, or created it, either naturally or artificially.

We are never working on the actual model of this world, but we are actually working on the model for the higher worlds resembling this world. Because if this world is only a replica of higher worlds, made in their image, then many space objects and phenomena influencing all models cannot be real in our world, but they come as default knowledge from the higher worlds, as part of this entire creation. These do not have to have a real counterpart here, but they can have only their concepts and images, to affect this world by default as it does in the higher worlds, without the actual object or phenomena present here. This is how this world can last as far as the lower orbit of Earth, with everything else illusory, not actually there, but only an image or concept, to make everything seem real, while this is how this world can be set in place on the firmament only as Earth, with all lights and images from the sky going around Earth once a day.

Because as a created replica of higher worlds, you can never know how far this world spans. While from all the accurate data that we have currently, it might not span further than the lower orbit of Earth. In this manner, the sun, the moon, and all stars that we see at night might be only illusionary images in the sky, made to make this world dense, stable, credible, appealing, thrilling, and seductive, while all natural laws present in this world are consistent with everything present in our higher realities. Therefore, our natural laws of the universe along with all accurate facts and concepts, and all accurate reasoning behind them are consistent with a universe as wide and as complex as the actual universe present within our higher realities.

Why having such a complexity of natural laws and details? Everything relates with the meaning of this world. Everything happening in this world is meant to generate knowledge through reasoning and interconnectivity, and therefore it is meant to find solutions and successful ideas addressing not only this world, but addressing all our higher realities. Because our world, with all humans within, has a cognitive and interconnective nature and meaning, and it is used in the higher worlds for thinking, experiencing, reasoning, and interconnectivity. Our world is the actual social media for the souls, while everything happening here models everything happening within higher worlds, since with all knowledge similar, everything remains correspondent from one world to another, while maintaining correspondence in all videogame worlds that you create, all novels that you write, and all reveries that you have.

More precisely, if this world is naturally created, as part of a higher mind, it has a cognitive meaning, used throughout reasoning. Similarly, you have countless of inner mind worlds right now within your own mind, helping you reason alongside this book, as you imagine all these concepts, people, circumstances, societies, and entire civilizations and realities alongside this book, while elaborating this entire book into a third level intelligent conception of the human reality, with all details minutely understood and minutely consistent.

However, if this world is artificially created and not naturally created, through mind and computer technology combined, then it is used to interconnect higher beings or souls within very appealing social media worlds, very similar to their own worlds, for increased relevance. In this case, this world is not part of the meanings of Life and of the wider world anymore, yet it is still part of Life and the wider world, since it is still created in a normal natural living common higher mind, by all souls coming in this world, as in a higher common reveries that all souls have while maintaining this world. In contrast, all inner worlds of the human cognition are part of Life and of the meanings of Life, but only if you use the human mind to fulfill normal human needs and meanings, not consensual ones, since the consensus stands apart from the meanings of Life, while going against Life altogether.

Study our world closely, to find a significant theme here, as this world focuses on specific historic people, circumstances, and achievements, while also focusing on drugs and tyranny excessively. The souls themselves focus on drugs and tyranny excessively, yet this world has a very important historic theme, current events and circumstances happening right now and right here on Earth, yet you cannot distinguish them anymore, since they are always present, and you cannot realize how important they are. While also offering art and entertainment simultaneously, through an entire virtual mode of living, or offering only distinct fulfillments, natural or artificial, whatever is harder to obtain in the higher worlds. Yet from the very large number of higher beings coming here on Earth to live their normal lives, it means that the higher worlds must be strongly constrained consensually. Too bad for them, but they are the ones making their own higher worlds consensual, artificial, and therefore constraining and unfulfilling.

While they bring the entire Consensual Matrix with them when they come down, making this world just as meaningless as theirs. While with the invisible kingdom working overtime to make everything possible down here and up there too, now this is the actual model for the human reality. Yet do not be too harsh with your judgment, because you take the Consensual Matrix with you throughout the Internet, videogames, and social media, otherwise you get in trouble if you fulfill your living, natural needs everywhere freely.

Is this world not large enough to cover the stars and galaxies that we see at night? Why does it have to be as large as the lower orbit of Earth? Because no one ever goes beyond the lower orbit of Earth anyway, and therefore as an artificially created reality, it does not even have to span further than the lower orbit of Earth. While it can always seem to be as large as the entire universe, just by displaying the multitude of lights on the night sky. Yet if you can ever find a genuine picture of Earth taken from anywhere in space, from the Moon or from anywhere else, then that is the size of this world. Just search the Internet and learn the truth for yourself, but hurry up, because they remove all pictures and videos from all space missions that seem too fake. Yet since all space records are fake and seem fake, hurry up to find them.

Currently, they are very careful not show pictures or videos that cannot remain consistent with the laws of physics in space or on other space objects, and therefore they do not show pictures or videos anymore, but only some numbers in a table or some spikes on a graph, proving all current missions of space exploration. This is what trillions of dollars buy you throughout all missions of space exploration, a spike on the graph.

Because it already becomes very suspicious not to have one single genuine picture of Earth taken from anywhere in space, even from the higher orbits of Earth, after all these missions of space exploration. You can study all current pictures of Earth taken from space throughout all decades of space exploration, since they are only seven, to find them showing Earth from a lower orbit or closer, altered and artificially constructed to seem as they are taken from a higher orbit of Earth or from space. Yet with thirty thousand artificial satellites in the orbit of Earth, all equipped with video cameras, where exactly are the pictures and videos that they take? Nowhere, while the artificial satellites themselves are missing. While in general, if you find pictures of Earth taken from space with excessively enlarged continents and peninsulas, they are taken from a lower orbit of Earth, and cannot include the entire Earth, since they are too close to Earth, still showing a round image of Earth, but only local, not of the entire Earth, with everything that they show excessively enlarged. Currently, there is still an image of Earth taken in an earlier mission of space exploration showing the Arab peninsula excessively enlarged, since it is not taken from space as it is claimed, but from a lower orbit of

Earth, from the space shuttle flying at high altitudes, from a high altitude balloon, or from a military airplane flying at a very high altitude.

There are no cameras on satellites, space shuttles, and space stations, and they never fly past the lower orbit of Earth, since there is where our spacetime continuum ends. Yet when you study everything closely, you notice how even our higher world ends with the lower orbit of their Earth, as it does in many higher worlds above, all made one in the image of another. Which means that this model of the human reality applies here and in several worlds above, up to our Creator himself.

There might not be any space out there to go to, or the current authorities never allow humans to go in space past the lower orbit of Earth. Because once humans go in space, then the entire human status changes in the wider world to space explorers, and there will be a different kind of life and civilization taking place down here and up there, wherever humans decide to be and travel.

Yet they can always fake space missions since movies and the entire world of entertainment can fake everything to seem very real and very credible, yet the entire credibility is only relative, because the human technology improves considerably from one decade to another, while everything that seemed real and credible several decades in the past, seems currently fake. You can even tell how everything was faked in the past, since you can fake everything similarly, while everything faked currently can still mislead you. This is why they do not show records from all missions of space exploration anymore, because in a matter of decades, you can spot everything as fake.

Not only pictures and videos are fake, but the entire setting, since it still uses the laws of physics on Earth, even while exploring the Moon or the empty space, displaying clearly the entire fakery. Yet you must be a physicist or a mathematician to notice it, while with all physicists in the Brotherhood, they already know that everything is fake, and they keep everything secret. Secrecy itself is a masonic virtue throughout the entire legal consensual Brotherhood, and everybody keeps it under all circumstances.

If there is no space beyond the lower orbit of Earth, then the entire model of the human reality differs from all consensual models of this world offered by science, with Earth fixed in the firmament of the matrix of our higher world, with all default images of space objects shown in the sky moving around the Earth once a day, and with all souls coming in this world as they please from the higher worlds for all addicted and tyrannical reasons, similar to how you play your videogames here on Earth. The souls even play the videogames here on Earth, since they never have enough. Yet since they do not use displays and controllers in the higher world, but a direct brain computer interface, everything seems real throughout the lower created worlds, as though you are in the higher worlds. While with souls having souls having souls throughout the higher worlds, the entire human existence is very complex, while making the actual human reality very complex, yet always different than the big bang theory, the theory of evolution, and the theory of relativity that the invisible kingdom promote.

This world spans as far as Earth and its lower orbit, it is at the center of its firmament, the rest is placed there as illusory images in the sky moving around Earth once a day, everything is made in this manner by our Creator residing only dozens of worlds above, he made this entire cluster of realities with all souls and all humans included while allowing some higher souls to create their own worlds below within this entire cluster of created realities, while this is actually creationism.

More precisely, our entire cluster of created realities is artificial in nature, it has other meanings apart from Life and the wider world, yet since the common higher mind holding it is natural and therefore part of Life and the wider world, this world and this entire cluster of created realities are always part of Life and the wider world, only apart from the meanings of Life and the wider world, going at times even against Life and the wider world.

However, as you study this world and this entire cluster of created realities closely, you find everything created at a very high existential resolution, as though our Creator intended to maintain intelligent developmental meanings in life and in the world, corresponding to all meanings of Life and of the wider world, which is a very high achievement, since it already made all past intelligent golden human ages possible. Yet with the Consensual Matrix at close proximity with our Creator, it infiltrated and compromised everything, turning his entire cluster of created realities consensual, while forcing our creator to destroy it repeatedly. This is the last world, and never again.

After this entire study, which cosmological model of this world seems more plausible? Creationism, so far. Yet could there be an actual Creator at the origin of this entire reality? Yes, since all realities have a creator, counting in zillions. Every time you have a daydream, play a videogame, watch a movie, or read a book, it has an author behind it, which is actually a creator. The creator of that videogame, which is actually an inner computer reality.

While as you study all realities formed, held, and maintained in this world, all have a creator. You cannot even have a reality without a creator, as these are always part of Life. We have mind realities and computer realities created here in this world, all having their own distinct creator. Right now, as you read this book and as you reason alongside this book, you elaborate it continuously in your conscious intelligent inner replica of the world spanning the cortex, which is your actual conscious mind, with you its creator. You have created your entire intelligent conscious inner mind reality throughout learning and throughout your entire experience, you are its own creator, and there is where you live your life as a conscious intelligence every time you reason, plan, feel, daydream, mental model, and elaborate. While all intelligences of all mind realities do the same, since all intelligences think and reason in a similar manner by the zillions within their own inner mind worlds, as they are the creators of their own inner cognitive worlds.

Creationism everywhere, in all worlds and realities of Life

or of the wider world. I refer to the Supreme Creator of the entire wider world as the Deity, as Intelligence, or as Life herself, since all these are omnipresent, omniscient, and omnipotent, while they are always a supreme perspective of each other, while I refer to the creator of this world and of our entire cluster of created realities as our Creator. As a reference, Life herself is the life of the Deity or Supreme Creator, the wider world is his physical body, and Intelligence or the Universal Mind is his actual mind.

It is still creationism, while in our particular case, we live in a reality created within a reality created within a reality about twenty times below, as this particular world is the hard bottom of this entire cluster of created realities, still part of Life and the wider world, yet also at their bottom. Our Creator is only the creator of this cluster of realities where the souls and the living human beings live, while our Creator lives in the reality above our cluster of created realities, which is the last natural reality of the wider world, with the entire Consensual Matrix at his proximity, as the Consensual Matrix reaches and controls most of Life and the wider world. As we study this entire cluster of created realities closely, we notice how our Creator created everything with the intention to fulfill Life and the wider world, not the Consensual Matrix.

However, all higher souls using this entire cluster of created realities obey consensual duties primarily, through the Consensual Matrix, while making all dark ages possible, along with all addictions and all tyranny. It is different throughout the golden ages, yet currently, it seems that our Creator cannot switch everything to intelligent human fulfillment and to entire golden ages anymore, since all higher souls coming here agree consensually to serve the Consensual Matrix, while compromising our entire cluster of created realities. Yet considering that most of the intelligent life of the wider world lives life in medieval dynasties one dark age after another continuously, the few intelligent golden ages of the distant human past are still a high achievement for our Creator and his entire creation.

The Human Reality

Our Creator seems to be human up there in his higher natural world, dozens of worlds above this one, relatively similar to us down here dozens of realities below. With all these worlds from there to here made one in the image of another in the image of another all the way to us, which is the last world. Not only that this is the last world, but it also seems to be the farthest in the past. So far in the past compared to the higher worlds above, that people cannot interconnect their conscious minds anymore, allowing them to form similar common worlds themselves. We still have videogame worlds and social media platforms here, but nothing else.

You can notice clearly how a big bang theory makes this entire reality random, spontaneous, mechanical, and meaningless, disregarding our Creator entirely, with the invisible kingdom implementing and maintaining the big bang theory instated, always disregarding the Creator himself and even Life and the Deity, while profiting continuously. These are the higher consensual intentions of all higher souls coming in this entire cluster of created realities, and if they are consensual, while most of them are, they compromise and corrupt everything, for drugs, tyranny, and servitude. Yet this world has a significant potential to develop everybody, high above vices, addictions, tyranny, servitude, and animal instincts as you find currently in this world.

As a reference, the particular existential resolution of this world can accommodate cognitive systems above the third intelligent level, up to the fifth level, which is an achievement for our Creator even here in this last world. Since this is why you always have receding psychic abilities in some or many people, because these should be possible in everybody once fully developed. Yet here come Einstein and Hubble with their embarrassing big bang, or Hoyle with his similarly embarrassing cosmological theory to explain the world.

Hoyle, since after the Second World War, we had two distinct models of the universe. One was Fred Hoyle's steady state model described above, and the other one was Lemaître's or Hubble's big bang theory, advocated and developed by

George Gamow first, with his associates Ralph Alpher and Robert Herman, further predicting the cosmic microwave background radiation. Hoyle coined the name big bang during a BBC radio broadcast in March 1949. Let us follow this line of reasoning now, since it is consistent with the expansion of the invisible kingdom as it took over the world.

In 1912, Vesto Slipher measured the first Doppler shift of galaxies, discovering that almost all nebulae were receding from Earth. He did not understand the cosmological implications at that time, since it was highly controversial if what they called nebulae were within or outside the Milky Way.

In 1922, Alexander Friedmann, a Russian cosmologist and mathematician derived the Friedmann equations from Einstein's equations of general relativity, showing that the universe might be expanding. This came in contrast to Einstein's old model of a static universe, which I described above.

In 1924, Edwin Hubble's measurement of the great distance to nebulae proved that they were other galaxies.

Georges Lemaître, the Belgian physicist and Roman Catholic priest, derived Friedmann's equations in 1927, stating that the inferred recession of the nebulae was due to the expansion of the universe.

In 1931, Lemaître stated that, if the universe expands, then the further in the past we are, the smaller the universe is, until we get to a finite time in the past when all the mass of the universe is concentrated into a single point, a "primeval atom," when the fabric of time and space came into existence.

Starting in 1924, Hubble developed a series of distance indicators, the forerunners of the cosmic distance ladder, using the 100-inch Hooker telescope at Mount Wilson Observatory. This had allowed him to estimate distances to galaxies whose redshifts had already been measured.

In 1929, Hubble discovered a correlation between distance and recession velocity, now known as Hubble's law, which I had described above. Lemaître had already shown that this was expected.

The Human Reality

In the 1920s and 1930s, almost every major cosmologist was embracing an eternally static universe, with several complaining that a beginning in time for the universe implied by the big bang brought religious concepts into physics, mostly since the inventor of the big bang theory, Monsignor Georges Lemaître was a Roman Catholic priest.

We note that at that time, the big bang theory was not part of the scientific consensus. Where was the scientific consensus, mostly when Einstein and Hubble were already supporting the big bang model and seemed to be the ones to lead and influence closely its development? The answer is that we are still in the old world order, under the old European aristocracy, while most of the people of the old world order had spiritual models of the universe, while explaining the entire spiritual paranormal world with all souls included. The world was still developing scientifically, socially, artistically, and spiritually under the old European aristocracy, mostly in the end, before the two world wars, while also seeming to diverge slowly away from the Consensual Matrix.

The two world wars had facilitated the new world order, the expansion of the invisible kingdom to cover the Western Civilization entirely, in all social domains, and through it, to cover most of the world. This is how, at the end of the second world war, only years later, miraculously, the entire world embraced a unique, universal idea not only in cosmology with the big bang theory, but in physics and science in general, in art, history, religion, politics, and sociology, forming a genuine global culture, because the paradigm had already flipped. Yet we notice a continuous nihilism ever since the new world order, with all art, literature, science, and the entire society contorted, oriented only towards profit, tyranny, and addictions, while matching entirely the invisible kingdom.

Ask anyone to explain the big bang, and they will tell you that the universe exploded long time in the past, and now it still expands naturally. Ask them why the entire world assumes so, and they will not know, or they will tell you about the redshift and the background radiation. Because not too many people are able to integrate the big bang idea in a larger model of society, or in a larger model of our civilization to show you clearly why this model of the universe had been embraced globally, unanimously, how it had been done, and which are the real forces directing all scientific efforts and resources towards the big bang, towards this redundant and very expensive global research project, with no other cosmological models to choose from, no alternatives to consider. Furthermore, they certainly do not research subjects as the big bang in all classified facilities as Area 51, since they do real physics there.

This is how we came to learn, think, and teach that the universe expands at incredible velocities and it had originated with the big explosion, the big bang itself, 13.8 billion years ago. While if you lived only decades ago, you had to believe that the universe is static. Yet if you lived a few centuries ago, you were forced to think and say that the Earth is flat and the universe revolves around it every day. Yet if we lived in the times of Eratosthenes and before, then the Earth was spherical, with all longitudes and latitudes on it, but with no turtles and elephants holding it. Yet there were turtles further in the past, and then elephants, with the golden egg around, or with everything in the golden egg, and with tears and ejaculation everywhere. Yet this world is distinct and unique, and it always remains unique and therefore it should have only one model describing it accurately, while ideologies and indoctrination can be anything ever while counting in tens of thousands, and while explaining everything in ten thousand manners.

What is the difference, when you are forced to believe one model of the universe or another, as these are simple beliefs but not accurate knowledge? Since this is why I persist to accept and use only accurate knowledge and not beliefs, assumptions, consensus, ideologies, and theories in my model of the human reality, regardless if these beliefs are at the base of all consensual human knowledge, since as we always notice, the consensual is not the real.

Because if I cannot find an accurate picture or video taken in space or on the Moon, but only artificially constructed pictures and videos showing transparent astronauts at times, or images taken in studios in ignorant manners by ignorant scientists, then I cannot accept these records to have been taken outside Earth. While after this entre time, if there are still no records from space and from the Moon, how can I even state that everything beyond the lower orbit of Earth actually exists objectively? They might certainly exist in the higher world where our Creator lives, or not, since I have nothing accurate and intelligent to state the truth. While with countless of fake records taken in space and on the Moon, there is nothing accurate and conclusive that I can state about this world past the lower orbit of Earth.

Because there are countless of pictures and videos that I get to see on the media, while they are not accurate but only fictional. There are space probes landing on moons, comets, and planets, while even those records are fake and inconclusive. When China sent its first probe to the moon, their video showed the artificial animation of an asteroid while it fell and burned in the atmosphere of the moon, with all the scientists involved laughing, shaking hands, and embracing each other, for their wonderful success. The atmosphere of the Moon! Similarly, the images taken from the international space station show an Earth always covered by clouds and water exactly as it looks from a lower orbit. Yet even when you study these, you see cloud patterns impossible to form by the laws of physics on Earth, but only by fiction, whatever those making them thought that looks more convincing. Study them closely, to see the same cloud formations covering casually most of the Earth, which is impossible naturally, since those types of clouds are only local on Earth.

The universe was denser and hotter in the past. In particular, the big bang model suggests that at some moment, all matter in the universe was contained in a single point, which is considered the beginning of the universe, 13.82 billion years ago. This is considered officially the age of the universe, yet this number always varies, while it was supposed to be unique, since all ages are unique.

Very shortly after the big bang, the universe cooled sufficiently, to allow the formation of subatomic particles, including protons, neutrons, and electrons. Giant clouds of these elements later coalesced through gravity to form stars and galaxies, and the heavier elements were synthesized either within stars, or during supernovae, or this is what they say.

What is the structure of the universe? What allows it to contract to an infinitesimally small point? What makes it explode? What makes it expand? What are the exact laws of the universe answering these questions? Does the universe behave exactly according to the big bang model? No.

The theory of general relativity describes spacetime by a metric, which determines the distances that separate nearby points. These points, which can be galaxies, stars, or other objects, are specified using a coordinate chart or "grid" that is laid down over all spacetime. The cosmological principle implies that the metric should be homogeneous and isotropic on large scales, which uniquely singles out the Friedmann–Lemaître–Robertson–Walker metric. This metric contains a scale factor, which describes how the size of the universe changes with time. This enables a coordinate system to be made, called comoving coordinate.

In this coordinate system, the grid expands along with the universe, and objects that are moving only due to the expansion of the universe remain at fixed points on the grid. While their comoving distance remains constant, the physical distance between two comoving points expands proportionally with the scale factor of the universe. This means that the big bang is not an explosion of matter moving outward to fill up an empty universe, but space itself expands with time everywhere, increasing the physical distance between two comoving points.

In 1964, Arno Penzias and Robert Wilson discovered the cosmic microwave background radiation. The radiation is consistent with an almost perfect black body spectrum in all directions. This spectrum had been redshifted by the big bang, and today it corresponds to 2.725 Kelvin. The cosmic microwave background radiation occurs shortly after recombination, the epoch when neutral hydrogen becomes stable. The discovery of the cosmic microwave background radiation had tipped the balance of evidence in favor of the big bang model.

Measurements of the redshift magnitude relation indicate that the expansion of the universe has been accelerating ever since the universe was about half its present age. To explain this acceleration, general relativity requires that much of the energy in the universe consists of a component with large negative pressure called dark energy. Dark energy solves numerous problems. Measurements of the cosmic microwave background indicate that the universe is very nearly spatially flat, and therefore according to general relativity, the universe must have almost exactly the critical density of mass-energy. The mass density of the universe can be measured from its gravitational clustering, and is found to have only about 30% of the critical density.

The exact nature and existence of dark energy remains a mystery. The universe consists of 73% dark energy, 23% dark matter, 4.6% regular matter, and less than 1% neutrinos. According to this theory, the energy density in matter decreases with the expansion of the universe, but the dark energy density remains constant as the universe expands. Therefore, matter made up a larger fraction of the total energy of the universe in the past than it does currently, but its fractional contribution will decrease in the far future, as dark energy becomes even more dominant.

During the 1970s and 1980s, various observations showed that there is not sufficient visible matter in the universe to account for the apparent strength of gravitational forces within and between galaxies. This led to the idea that up to 90% of the matter in the universe is dark matter that does not emit light or interact with normal baryonic matter. Additionally, the assumption that the universe is mostly normal matter led to predictions strongly inconsistent with observations. In particular, the universe today contains far less deuterium than can be accounted for without dark matter.

There are still three outstanding issues with the big bang theory: the horizon problem, the flatness problem, and the magnetic monopole problem.

This was, in short, the big bang model of the universe. We note that the theory behind this model is inconsistent with astronomical observations, forcing physicists to adjust constantly their constants and to invent major states and components of the universe, as inflation, dark matter, and dark energy, only to explain unexpected observations as the accelerated expansion of the universe, or the need for very large gravitational forces everywhere coming from dark mass, in order to keep the universe together. This entire model of the universe had started from the wrong assumption that the redshift of distant galaxies is caused by them moving away from us at relativistic velocities, made one century ago, when physics and astronomy were not too advanced, and when computers were inexistent.

I am including here a very simple timeline of cosmological theories and discoveries to summarize the development of humanity's understanding of the universe and therefore of the human reality, exactly as it is agreed upon by the scientific consensus and by all current authorities controlling it, and with the exact dates and knowledge agreed upon. For the next chapter, we build our own model of the universe and of the human reality from scratch, an intelligent conceptual model, starting directly with the definition of the universe. For now, let us see the timeline of everything studied so far.

16th century BC - Mesopotamian cosmology has a flat, circular Earth, enclosed in a cosmic ocean.

6th century BC - The Babylonian world map shows the Earth surrounded by the cosmic ocean, floating on water and overarched by the solid vault of the firmament to which are fastened the stars.

4th century BC - Aristotle proposes an Earth-centered

universe, in which the Earth is stationary and the universe finite in space and infinite in time.

3rd century BC - Aristarchus of Samos proposes a Sun-centered universe.

2nd century AD - Ptolemy proposes an Earth-centered universe, with the Sun, moon, and visible planets revolving around the Earth.

964 - Abd al-Rahman al-Sufi (Azophi), a Persian astronomer, makes the first recorded observations of the Andromeda Galaxy and the Large Magellanic Cloud.

15th century - Ali Qushji provides empirical evidence for the Earth's rotation on its axis and rejects the stationary Earth theories of Aristotle and Ptolemy

15th-16th centuries - Nilakantha Socanaji and Tycho Brahe propose a universe in which the planets orbit the Sun and the Sun orbits the Earth, known as the Tychonic system

1543 - Nicolaus Copernicus publishes his heliocentric universe.

1576 - Thomas Digges brings the idea of the star-filled unbounded space.

1584 - Giordano Bruno creates the model of an infinite, static universe with infinite stars and spatial voids. All stars can have planets bearing life while the Sun is just an ordinary star.

1610 - Johannes Kepler claims a finite universe, since the sky is dark.

1687 - Sir Isaac Newton creates the physical laws to describe the large-scale motion throughout the universe.

1755 - Immanuel Kant asserts that the nebulae are galaxies outside Milky Way Galaxy.

1791 - Erasmus Darwin imagines the first description of a cyclical expanding and contracting universe in his poem The Economy of Vegetation.

1905 - Albert Einstein publishes the special theory of relativity, stating that space and time are not separate continua, but relative.

1915 - Albert Einstein publishes the general theory of relativity stating that high energy warps spacetime.

1917 - Willem de Sitter derives an isotropic static cosmology with a cosmological constant, as well as an empty expanding cosmology with a cosmological constant, termed a de Sitter universe

1922 - Vesto Slipher summarizes his findings on the spiral nebulae's systematic redshifts

1922 - Alexander Friedmann finds a solution to the Einstein field equations which suggests a general expansion of space

1923 - Edwin Hubble measures distances to a few nearby galaxies. These distances place them far outside our Milky Way, and implies that fainter galaxies are much more distant, and the universe is composed of many thousands of galaxies.

1927 - Georges Lemaître states the creation event of an expanding universe governed by the Einstein field equations. He predicts the distance - redshift relation from the solutions to the Einstein equations.

1929 - Edwin Hubble demonstrates the linear redshift-distance relation pinpointing the expansion of the universe.

1933 - Fritz Zwicky shows that the Coma cluster of galaxies contains large amounts of dark matter.

1934 - Georges Lemaître interprets the cosmological constant as due to a vacuum energy with an unusual perfect fluid equation of state.

1948 - Hermann Bondi, Thomas Gold, and Fred Hoyle propose the steady state cosmology.

1948 - George Gamow predicts the existence of the cosmic microwave background radiation by considering the behavior of supreme radiation in an expanding universe.

1950 - Fred Hoyle coins the term "big bang" to highlight the difference with his steady state model.

1965 - Martin Rees and Dennis Sciama analyze quasar source count data and discover that the quasar density increases with redshift.

1965 - Arno Penzias and Robert Wilson discover the 2.7 K microwave background radiation.

1966 - Stephen Hawking and George Ellis show that any

plausible general relativistic cosmology is singular

1967 - Robert Wagoner, William Fowler, and Fred Hoyle show that the hot big bang predicts the correct deuterium and lithium abundances

1970 - Vera Rubin and Kent Ford measure spiral galaxy rotation curves at large radii, showing evidence for substantial amounts of dark matter.

1976 - Alex Shlyakhter uses samarium ratios from the Oklo prehistoric natural nuclear fission reactor in Gabon to show that laws of physics are unchanged for billions of years.

1980 - Alan Guth and Alexei Starobinsky independently propose the inflationary big bang universe as a possible solution to the horizon and flatness problems.

1983 - 1987 - The first large computer simulations of cosmic structure formation are run by Davis, Efstathiou, Frenk and White. The results show that cold dark matter produces a reasonable match to observations, but hot dark matter does not.

1988 - Measurements of galaxy large-scale flows provide evidence for the Great Attractor.

1990 - Results of COBE mission confirm the cosmic microwave background radiation has a precise blackbody spectrum, eliminating the possibility of an integrated starlight model proposed for the background by the steady state model.

All these are erroneous, and they should have spent all money to feed the world.

8 ACCURATE MODEL OF THE HUMAN REALITY

What is the universe? Do a simple research on any encyclopedia, to see that the definition of the universe relates only to the big bang theory. The universe itself becomes the big bang theory by the year. Who is interested so much in the big bang theory? The interest is not particularly in the big bang theory, but in its limited definition. The big bang model makes our universe only 13.8 billion years old, which is younger than the human race itself, with most of our genetic material being even older. The big bang closes another door of perception to our past, our wonderful history spanning solar systems and probably galaxies. The big bang gives us a feeling of short existence, of irrelevance throughout life, and it truncates our eternal life into the limited today and a limited ending life. Why inducing amnesia to an entire intelligent race? In order to take something from us, and then keep it endlessly. The problem when someone or something takes something from us against our will is that we always manage to find a way to take it back, no matter how weak or how disadvantaged we are. Yet with amnesia, we can never claim and get back what is ours, since we cannot remember it.

The big bang model is also very unstable, and this adds to the feeling of incertitude that we experience. Give it a little more dark matter in one of its equations, and the universes collapses in another 13.8 billion years. Take away some dark matter from another equation, and the universe expands endlessly, making the mass density so low, that no stars and planets will ever form. Another characteristic of the big bang model is that it includes only the plane of existence that we call this world, and it cannot explain anything else, none of the planes of existence that we are familiar with and inhabit frequently throughout life. It could be from the pure ignorance of those who had created or demanded it, or it could be simply to shut down another door of perception leading to the entire wider world, the wider world including other planes of existence. Since with an inadequate model of the universe, knowledge is kept in distinct modules, less elaborated, making reasoning more tedious, even impossible. This is how physics studies the structure and behavior of all objects and phenomena in the objective world, psychology studies the behavioral and cognitive processes related to human beings, nothing studies dreams and astral planes, art studies only painting and photography but not art itself, while computer science studies only computers and computer networks. This entire encapsulated data is not interconnected as it should be, and it has nothing in common in this manner, even though it is part of the same universe, therefore obeying the same natural laws of the universe. If you still wonder what these encapsulated modules of knowledge should have in common, then you are the perfect citizen, and everybody loves you, in an entire undeveloped exploited world.

The big bang theory is incapable to explain anything in this world, but only what happened three seconds after its own explosion or what happened when the cosmic microwave background radiation was produced. It seems that all research is done only to patch up and repair this cosmological model and to prove continuously its accuracy, but not to explain the universe we inhabit daily in its entire diversity. A model of the universe should be the genuine model of everything, a genuine comprehensive model explaining the entire universe continuously, the feather of a bird, the human reasoning, positive social interaction, human ideas, centrifugal forces, and even all fluctuation of the price of rice in Bangkok, yet the big bang explains only two phenomena, the redshift of distant galaxies and the background radiation. An intelligent conceptual model of the universe should include and interconnect all modules fragmented by science, since everything is part of this universe. While without understanding the interconnectivity itself, you cannot understand the universe.

Furthermore, since everything happens in this world, your entire understanding of this world must help you detect, understand, and predict everything happening everywhere, all lines and lifelines of causality, because this world holds and makes possible everything, while all knowledge of this world and of all other worlds and realities must help you perceive, understand, elaborate, master, and predict everything ever.

More precisely, you knowledge of this world and of all worlds and realities along with all natural laws making them possible should always be at the base of your intelligent cognition in an intelligent accurate manner. How can you ever reason intelligently continuously through the big bang theory? It is done in this manner deliberately, in order for you to remain ignorant, incapable, vulnerable, and therefore exploitable, because tyrants will use everything to exploit you, your loved ones, and the entire world even indefinitely.

This is why we need to be careful when we start our model of the universe, since it is rationally intelligent and based on accurate facts, as it relies strongly on its definition, structure, and laws, with everything from the human reality based directly on it, sustaining all events, information, objects, subjects, and interconnectivity in the wider world and in all its realities. In contrast with the current model of the universe stated by the current science, that holds only two phenomena and only consensually, the redshift and the background radiation.

Because the accurate model of the universe must sustain from its base Existence itself, along with its spacetime continuum, which hold themselves all natural laws of the universe, including classical physics, relativity, mathematics, all events, objects, subjects, and intelligences, along with all concepts and definitions of all these, along with their continuous interconnectivity throughout all lines and lifelines of causality flawlessly defined and sustained by Existence and by the spacetime continuum throughout all worlds and realities of the wider world. This is significantly more than the two theoretical phenomena that the current big bang theory is capable to hold, the background radiation and the redshift of distant stars. How exactly do the background radiation and the redshift of distant galaxies feed the world, and how much do these cost?

We also notice that we do not aim only for the model of this universe, which is this world, but we aim to model the entire wider world, which is the universe, or Life, or Intelligence, or Interconnectivity. We model all these throughout this book series "Human," while now we model only this universe, or this world, along with the wider world, since all realities of the wider world are similarly modeled and defined, from the inner cognitive realities of your own daydreams and mental models to all videogame realities in the world, and to all worlds and realities of the wider world, inner and higher, including this world, called Earth or the universe.

This is why, if we state the definition of the universe erroneously or crooked from the beginning, we can end up with a crooked universe in all results. I do not invent the definition of the universe here, but I take it directly from our official sources the way they interpret and use the word 'universe' currently, while I consider its etymology, which helps us understand how the universe was understood ages ago, when the word 'universe' was first needed, created, and used. Let us start with the etymology of the word 'Universe'.

The word 'universe' comes from Latin and Greek. 'Uni' means one, or a single one, implying a single universe. 'Verse' in Latin might imply the notion of movement, it might mean one turn, as in one move, one turn of the game or one turn of the sphere. It is one move defining everything happening in the universe. In this manner, 'uni-verse' means a one, complete, distinct entireness and movement, which is the One. This further implies that everything happening in the universe is consistent, interrelated, and interconnected, creating a distinct, entire, harmonious movement. This is the Natural Law of Interconnection, which further derives the Natural Law of Attraction. This means that our universe is a single, distinct, consistent oneness including everything, happening everywhere, every time, endlessly. Simultaneously, everything happening in the universe is consistent, interconnected, it obeys the same laws, and it influences the entire universe, even as a whole.

More precisely, the universe is everything that can ever be objectively, and as far and as long as its spacetime continuum can hold it objectively. This is the actual definition of a reality, any reality, while these count in zillions throughout the wider world.

All realities have their own natural laws holding and determining everything within, all beings, objects, meanings, concepts, events, and lines of causality and lifelines of causality. The current natural laws of the universe as they are stated here seem empirical, yet they seem to be the actual natural laws of this world and of the worlds above within our cluster of created realities, also used in this world for consistency and relevance. Furthermore, our Creator used most of the natural laws of the wider world while creating our entire cluster of created realities, in order to match the meanings of all souls and all humans with the meanings of Life and of the wider world.

These natural laws of the universe set in place the natural laws meant to describe the interaction between the field, the spacetime continuum, matter, and radiation, and they include mathematics, classical physics, electricity and magnetism, electrodynamics, relativity, along with all derived knowledge coming directly from these. I refer to these as accurate knowledge, distinguishing them from beliefs, dogma, and ideologies, which are consensual knowledge.

More precisely, all accurate knowledge relate directly to all natural laws of the universe, which relate directly to the universe itself and its existence and continuum. While all consensual knowledge is only agreed to be true, making it only consensually true, including all theories, speculations, beliefs, ideologies, statutes, jurisdictions, laws, regulations, corporations, districts, and politics. Since this is why we have to model the human reality distinctly into the real natural living human reality on one side, and the consensual human reality on the other. Reality itself is unique, only one, yet consensually, people can agree on anything, making reality everything else, but only for them, whatever they decide. Which should never affect you if you never agree with them, yet it does, continuously, just look around to see it for yourself, because in an undeveloped consensual world, everybody makes agreements against you, mostly to exploit and harass you.

The natural laws of the universe affect and determine closely all lines and lifelines of causality, everything happening in the world. All natural laws of the universe consisting mathematics and classical physics determine closely all lines of causality everywhere around you, from the motion of everything around to optics, vibrations, images, heat dispersion, rotations, and more. Yet whenever life is involved, it becomes harder to predict everything anymore, because all living beings with humans included behave in any predictable and unpredictable manner, causing and determining everything lively and naturally. While these are not predictable lines of causality anymore, but genuine lifelines of causality. Good luck defining these, since everything looks like soap opera in the living world throughout countless of lifelines of causality spreading everywhere while affecting everybody directly and implicitly.

While we have to be able to define all these throughout our current accurate model of the human reality, this accurate model of the universe as we attempt to find here, which is just as dense and complex as the entire world. Yet can we ever model it? We certainly can, better than the big bang theory, but we have to trace some limitations. We can only trace the main details of the model of the universe here, because it is too dense and too complex. Furthermore, we cannot match the existential resolution of this world in our model from this book, because we do so from within this world, making it impossible. You might be able to model the entire "Horizon Zero Dawn" videogame reality, modeling and explaining everything in details, as you do so from within this world, which stands above the videogame reality. Yet even so, notice how some of the natural laws of the universe have to be used throughout many videogame realities in order to make them credible and correspondent with this world. Since this is why they use physics in the videogame reality called "GTA5," in order for all cars to feel and behave normally and naturally there, just as cars do here in our world. Which is tedious and impressive, yet with a revenue in billions from "GTA5," it must be worth it.

Additionally, we want our model of this world to match and remain correspondent with all realities of the wider world. Which means that we have to model not only this world, but a general, most common reality just as well, correspondent to all realities of the wider world, zillions in number, higher and lower, while we have to do so from within this world, yet for the entire wider world.

We want to model the human reality, which is a third level intelligent conception corresponding to third level living human beings, which is possible, always maintaining compatibility. Yet then we want to define all realities of the wider world, which can be from the first consensual level of the entire Consensual Matrix, to the tenth supreme level, which is the level of the entire wider world, and we cannot handle anything past the third intelligent level. This world is of the third intelligent human level, matching humans by nature, as it is only a created reality, or this is what all the souls state, as it was created for humans, at least at the third intelligent level.

Otherwise, none of these third level concepts worked anymore, and you could never reason and behave as minutely and as intelligently here as you do in the higher worlds. Since this is what makes this world so dense, so stable, so credible, so involving, and so seductive.

There is one more limitation related more to these words and not exactly to humans, because these words from these paragraphs allow only a first level algorithmic communication, never being capable to hold any mental model above the first algorithmic level. This matches perfectly everything consensual, ideological, and juridical, since these are of the first level, but not anything alive, since life starts with the second animal level, with humans and human realities at the third intelligent conceptual level. The actual mental model of the human reality is in your mind right now, since you are the one making the mental model while reasoning alongside this book continuously. Which means that you have started the mental model of the human reality at the beginning of this book and long before, because everything that you learn throughout life is part of the human reality, natural and consensual. Since this is why an accurate model of the universe is very helpful for you, as it should stand at the base of everything that you know, learn, and experience throughout life, because you live your life in this world, and you should be able to know, master, and model it entirely. Otherwise, any consensual substitute for the accurate model of the universe renders you ignorant, incapable, weak, vulnerable, unsuccessful, and eventually exploited, as it is the case with many people in this world.

We have always distinguished throughout this book between consensual knowledge and accurate knowledge. Consensual knowledge is true only within the specific ideologies, jurisdictions, and consensual matrices holding and promoting them, as science, various jurisdictions, various nations since these are currently corporations and jurisdictions in themselves, and entire ideologies. While whatever the people of these ideologies and jurisdictions decide to accept through their consensus remains consensually valid, but only within their own ideologies and jurisdictions. In contrast, accurate knowledge is accurate everywhere within this world, which is the universe, including Earth. Furthermore, accurate knowledge, being accurate everywhere in the universe including on Earth, always remains consistent with all natural laws and natural knowledge defined as true and accurate through the specific natural laws and natural knowledge used to form, hold, and maintain this world, which are the natural laws of this universe.

Which means that, the unique, accurate model of the universe that we persist to create right now has to remain accurate for all the natural, living realities of the wider world including this world, while remaining accurate through the same model, for all consensual realities of the Consensual Matrix, including all its districts, courts, codes of law, jurisdictions, and corporations. While remaining accurate in a similar manner for all videogame digital worlds from all computers of the world, and for all daydream and inner mind realities that you can ever imagine and undergo yourself along with everyone in the world, counting in zillions. This is significantly better than explaining only two consensual phenomena, the redshift and the background radiation, as the big bang theory does today.

All realities of the wider world are formed or created naturally, artificially, or consensually, except for the ultimate reality at the top, which always exists. You can always form, create and maintain any reality from its higher reality above, only through a matrix or a buffer between the two realities, and under very strict circumstances.

This is the case with all realities that are made or created naturally and artificially, since all realities are formed, held, and maintained within a matrix that is formed, held, and maintained in the higher reality, where all inner realities are made or created. As always seen, realities have to be made, formed, or created either naturally, consensually, or artificially, from within their higher realities, since you cannot make new realities from within your own reality. Fictional characters cannot write their own book in their own fictional reality, with them and the rest of the characters within. While you cannot imagine this world yourself, as you live your life within this world, but you can daydream and imagine only within your own inner mind worlds, which are your own inner realities, created by you from here from this world, always standing above them. Similarly, Mario from your videogame cannot have made his own videogame called Mario on his own with him inside, because all realties are made, held, and maintained in and by their own higher reality, in a specifically encoded matrix or buffer. There are always one or more creators involved up there in the upper reality throughout their creation, since all realities have their own creators, regardless if they are naturally created, artificially created, or consensually created.

As a reference, all intelligent inner mind realities that you create right now while reading this book are naturally created, you are their creator naturally, and therefore they are part of Life and the wider world, while they are also part of the meanings of Life and of the wider world, since you fulfill your natural intelligent developmental need for learning and accurate knowledge while reading this book. All intelligent inner mind realities that you create right now are part of the wider world, correspondently similar to all natural realities of the wider world. In contrast with this world and with this entire cluster of created realities made by our Creator, because this world and this entire cluster of realities are not part of the meanings of Life and of the wider world, while they are still part of Life and the wider world.

Currently, we have a multitude of distinct types of realities to consider, as natural realities, mind realities, consensual realities, digital realities, fictional realities, artificial realities, mythical realities, social realities, astral realities, etheric realities, cellular realities, ionic realities, cognitive realities, and higher realities. We can always distinguish between the natural and the artificial realities. Artificial realities are consensual, digital, or fictional realities. I refer to natural realities as living realities, since they are part of Life and the wider world. The multitude of mind realities from your mind, reasoning, and imagination are natural, living, intelligent realities, since they are part of you, and you are a natural living human being, part of Life. Your own intelligences within your mind are distinct, independent living beings, and at their own inner level, they are made of living beings themselves, all living a normal life within the inner worlds and inner inner worlds of your mind, in a relatively similar manner as you live here in this world. Therefore, if this world is natural and therefore alive and part of Life, it is a biological, cognitive reality in itself, as part of a higher being or group of higher beings, more likely part of their inner mind worlds, or common inner mind worlds.

How exactly do beings make, form, or create their realities, natural and artificial? They imagine them entirely throughout beautiful reveries, while in this manner creating all possible inner mind realities. These realities are part of Life and the wider world, since you and your entire mind are naturally part of Life and the wider world, yet the inner mind worlds that you create while daydreaming are not part of the meanings of Life and the wider world, because you do not daydream within them in order to fulfill your intelligent human needs as part of Life and the wider world, but you daydream for any other reason, whatever you daydream.

Furthermore, you can create digital artificial realities by writing the necessary computer code for the multitude of videogames that you play and therefore experience and inhabit, yet these are not natural realities part of Life and the wider world, but only artificial realities, apart from Life and the wider world. However, if you use your computer programs to fulfill your intelligent human needs, all digital created realities become part of the meanings of Life and of the wider world, and it still counts.

Furthermore, while creating entire consensual realities, as entire ideologies, entire districts, or entire jurisdictions, people can form the necessary sets of agreements, laws, and beliefs as consensual buffers or consensual matrices capable to hold the actual consensual realities that they hold, which is the case with all districts, jurisdictions, and ideologies. Consensual realities do not exist, while the Existence of Life and of the wider world cannot define them, yet the people themselves define them consensually, in a fiat manner, never there. Only for them these consensual realities have validity, while they never have accuracy, not even for them.

As you study closely all consensual realities, all ideologies, and all jurisdictions, you notice how they maintain consensual agreements on everything that is not the case in life and in the real world, since if it was already the case, people never had to agree on them, because they were already present and already the case in life and in the real world. In this manner, all consensual realities are never part of Life and of the real world, while many times going against life and the real world.

You are familiar with the legalese language of all jurisdictions, since everything is contorted consensually throughout all jurisdictions, including language itself. All jurisdictions have their own statutes, where they define all worlds, all concepts, and everything else in any inaccurate manner they choose, while in this manner, always standing apart from life and from the real world, and from the meanings of life and of the real world, always against humans, humanity, life, this world, and the wider world.

All districts and jurisdictions must maintain legality among themselves otherwise they harm themselves and are dissolved, while through this comprehensive consensus, they form together the Consensual Matrix, which spans most of the wider world, while always harming Life and the wider world.

The most common created realities of Life and of the wider world are the actual mind realities, which are also the most common created realities here in this world, giving to the entire wider world its comprehensive cognitive characteristic, and this is why the wider world is also referred to as the Universal Mind by spirituality. All mind worlds are part of Life and the wider world, while they are also part of the meanings of Life and of the wider world, but only if the living beings themselves are not consensual, part of the Consensual Matrix. People learn and experience throughout lifetime, while forming their own inner replica of the world in their mind, which is a natural reality of the wider world, holding all their memories, feelings, and understandings of the world, where they can live, reason, mental model, and daydream as a conscious intelligence or as an intelligent inner self.

There is a difference between the conscious intelligence and the intelligent inner self. The intelligent inner self from your intelligent inner conscious mind world consists of everything that you perceive, know, learn, and understand about yourself, since your intelligent inner self is your correspondent replica of yourself from the outside world, living in your intelligent inner mind world as though it is outside in the real world. Your conscious intelligence is different, since your conscious intelligence lives above your intelligent mind reality yet still in your mind, while it uses the intelligent inner self from the intelligent inner mind world as an avatar in order to access, reason intelligently, and live your life within the intelligent inner mind world.

As a reference, your intelligent inner mind world is the only intelligent mind world of the human mind, since only the cortex is capable to offer intelligent cognition. While you must always have a self of yours already in a world or in a reality in order for you to be there as that particular self. The souls refer to all selves as avatars, while the word avatar is among the oldest worlds used by humans here in this world. The souls themselves cannot be in other worlds and realities if they do not have a self or avatar already there, and this is why they come into humans, while using them as selves or avatars in order to be here in this world.

More precisely, all souls come into the conscious intelligence first, while as a conscious intelligence, they come into the inner selves to reason, feel, and daydream, while as any inner self, intuitive or intelligent, they come into the physical body from this world, as actual human beings. Yet the souls can be all these simultaneously, since they are in this manner one through another through another, allowing them to reason within the inner mind worlds while moving around in the outside world simultaneously.

You must always have a self or avatar to be in another reality, in your inner mind reality, as an inner self. In this manner, you live your life in your own inner conscious mind as an inner self, while reasoning, feeling, needing, and daydreaming. Similarly, you are your own videogame character when you play videogames, which is your own avatar from your videogame, while only through your videogame character, or only as your videogame character, you can be in your videogame world. Similarly, you are your own consensual corporation in order to be and therefore exist consensually within the specific jurisdiction of society where you always are throughout life at work, in court, at the lodge, or at the office.

How exactly do realities exist? Realities exist subjectively from the perspective of their higher realities, while these higher realities can be subjective in nature, for all higher realities above, since there are always entire chains of realities defined by Existence throughout entire lifelines of causality and lifelines of existence, involving all living beings, while this is Life, the wider world, Interconnectivity, or Intelligence, since it is the same.

This specific model of a reality that we make in this chapter and throughout this book is significantly more complex than anything offered by science, because science cannot leave this world throughout all its studies of the universe. This is how the current science ends up with an erroneous study, and this is how it needs an entire scientific consensus to validate it consensually, not accurately. The current science is based only on a scientific consensus validating itself, which means agreement, making it consensual entirely, and therefore not accurate. It is not real, nothing there, but only consensual, fiat.

It is similar with psychology, since psychology never leaves this objective world in order to study and describe the human mind, while the human mind is subjective in nature, being an inner reality in itself, never part of the real outside world, filled with a multitude of inner mind realities itself in very large numbers. Furthermore, science never leaves Earth throughout all its research, even while studying the universe, since this is why there are no space records available.

Science tends to remain within the consensual part of the world, maintaining the Earth coordinate system and the Earth mentality continuously, because the current science is actually the science of the Consensual Matrix on Earth, ending up making a model of the consensual world, and not of the natural living world. This is why science ignores altogether all natural, living ages of Earth, as the Age of Aries, demonizing it and ending up disregarding everything happening before a few thousand years ago. While even when it studies older cultures and older nations, it studies only the Sumerians, Babylonians, and Mesopotamians, and later on the Greeks, the Romans, and the entire Western Civilization, since only these were and still are in the Consensual Matrix. More precisely, the Consensual Matrix never steps outside itself.

While this makes the difference between the living and the consensual throughout the world, between the living and the consensual existence throughout your own life and existence here on Earth, between the consensual truth that you learn in society and the accurate truth that you have to study and learn yourself because no one will offer it to you, between the consensual human reality and the actual natural living human reality, between you the living human being and the consensual corporation that you are in society, at school, at work, on the road, in your city and nation, and everywhere else in the Consensual Matrix.

How are realities made, held, and maintained? Now that we are capable to consider other realities besides this world, even higher realities, holding their own created realities in a matrix, can we at least model the big bang? Because we can certainly imagine in our mind the big spontaneous explosion forming this world, and now we can hold all possible points of reference in the higher reality and in our world. Yet the big bang is still impossible even when we consider in our model higher realities, which is more than what science does, because the spacetime continuum of our higher reality does not coincide with our spacetime continuum to make the big bang possible. Realities are not formed by clipping away spacetime continuum from the upper reality in order to make the lower realities the way you manufacture shirts, but the upper realities form and maintain an abstract encoded matrix that holds and maintains the inner realities below.

While you imagine this big bang explosion right now in your mind, you create an actual inner reality in your mind through the big bang that you imagine yourself, which means that it is possible to create any reality through any big bang, but only when you imagine it yourself. More precisely, you cannot use a ruler from our world to measure the created big bang from your imagination in order to define and therefore model it, since all spacetime continuums span only their own realities, never those above or below. Because all inner realities are not exactly created through these imaginary explosions, but directly, through thoughts or computer code here in our world, or through direct mind interconnectivity in the higher worlds, if this ability is ever available. Yet they use technology to link minds by the billions in the higher worlds, and this is how they hold and maintain our world, still making it part of Life and the wider world.

The current authorities use the big bang theory only to bypass the Creator of this world, ignoring it continuously, while the invisible kingdom refuse to accept the Creator of this world altogether.

All realities need a matrix to hold them, since they cannot be held directly in their upper realities, because realities cannot intersect each other. Computer videogame realities are held in a specifically encoded digital matrix formed, held, and maintained by the multitude of little switches on the motherboard. The motherboard, the electricity, and all switches are material, objective, and therefore part of this world, but the specific artificially intelligent modulated encoding of the opening and closing of all switches is subjective, and therefore it is not part of this world anymore, but it is already part of the inner digital computer world. This is the computer matrix, capable to hold the continuum of the computer inner subjective reality, which is actually the computer binary code, or the computer language, the ones and the zeros encoded. While through this specific intelligent code, we are already within the inner, computer reality, capable now to hold the operating platform, which holds your Linux, Mac, or Windows operating systems, which hold your videogame, which is the actual inner, subjective videogame reality, and this is how you create a reality. While if you do so yourself, you are the creator yourself, with all your videogame characters venerating you continuously since you are their creator.

It is the same with all inner realities of Life and the wider world, held by all living beings and intelligences of the wider world, only that the matrix that these form, create, and give birth to is not artificial and not digital as it is the case with computers, but it is natural, intelligent, and alive, based on the encoded fluctuation of two distinct states of the electromagnetic field, electric and magnetic. This is the ionic form of life, standing at the base of the molecular form of life, standing at the base of the cellular form of life, standing at the base of the organic form of life, standing at the base of the social class of life, yet only if society is human but not consensual. Consensual means dead, not living, neither real, nor naturally intelligent.

All consensual realities are formed by the specific agreement standing at their base, capable in this manner to define directly, by agreement, all the necessary consensual matrices, continuums, platforms, courts, districts, laws, and corporations. This is why they make a court around you to judge you, only to take you symbolically or consensually away from the real living world, into that specific consensual reality called court, jurisdiction, or district, in order to judge you there in any manner they want, because that is their world or their territory, and they have complete authority over you when you are there.

Furthermore, you cannot be in any court, jurisdiction, or district as a living human being, because all living human beings live and exist in the real world, but only as a corporation, which is a consensual invention made in order to change your nature consensually from a living human being to a consensual corporation, since only corporations can exist consensually in all courts, districts, jurisdictions, and in all ideologies. Jurisdictions themselves are ideologies, they are legal ideologies or juridical ideologies.

All inner realities must have a specific natural or artificial encoding in their matrix, as laws, agreements, and various characteristics defining them. Videogame realities are digital realities, while all lines of computer code are their higher laws defining them in all details, becoming now their own natural laws. However, from within these inner realities, you cannot have access to these lines of codes yourself, as you can perceive and understand them in the manner that they transpose in the inner reality as natural laws, therefore affecting you directly, since they stand at your base. This is how, for our world, there is a significant difference between the specific higher laws, agreements, and characteristics defining and determining this world here, within, yet we can always perceive these as our natural laws of the universe. We are lucky to have them at the third intelligent level, matching us as living human beings, otherwise we could never reason as we do now, and we had to remain indoctrinated continuously at the first consensual level.

It is such a significant difference between the natural laws of the universe studied now in these few paragraphs, the specific laws of mathematics, classical physics, relativity, and electrodynamics that you know well, the spacetime continuum and the electromagnetic field holding everything defined by these, and the specific supreme higher laws defining everything from our higher realities to be exactly as it is defined here within. Because our higher laws define us from the other side of our matrix holding our world, while we perceive them from here from this side, within this world, as the natural laws of the universe. The natural law of Mentalism, the natural law of Correspondence, the natural law of Vibration, the natural law of Polarity, the natural law of Rhythm, the natural law of Causality, and the natural law of Gender.

These are the natural laws of the universe as the souls state them or as the people of this world have managed to identify them in this world, yet it seems that these are the natural laws of their own higher worlds, applied now through correspondence in our world, since all our natural laws apply to all worlds and realities made by our Creator, to the entire cluster of created realities. While the other worlds and realities of the wider world have their own natural laws defining them, yet it seems that our Creator integrated most of them at the base of all worlds of his entire cluster of created realities, in order to maintain concordance between the meanings of Life and all meanings of all souls and human beings from the entire cluster of created realities.

Furthermore, at the tenth supreme level, which is the ultimate level of Life, Intelligence, Interconnectivity, the wider world, and the Supreme Being, their own supreme laws define them and their own supreme world minutely. Yet since all realities of the wider world are correspondent with their own inner subjective worlds and realities, these supreme laws define these implicitly and correspondently, depending on cases, with some supreme laws part of our world and part of our own natural laws just as well. The most significant supreme laws of Life and the wider world are: the supreme law of Mentalism, the supreme law of Correspondence, the supreme law of Vibration, the supreme law of Causality, the supreme law of Meaning, the supreme law of Development, and the supreme law of Harmony.

You can also perceive our higher laws in other forms, as through mathematics, classical physics, and electrodynamics, or through literary and artistic characteristics, various spiritual and religious laws, biological laws, interconnective laws, social laws, and psychological laws, since these are correspondent, as they are only of a different language or encoding.

The official definition of the universe is: 'the universe is everything that exists.' While after this entire study, we can modify the definition of the universe slightly: the universe is everything that exists objectively, because existence is subjective, objective, and highjective in a relative manner. Everything that exists subjectively does so only in the multitude of our inner realities, including our mind realities and computer realities. Everything that exists highjectively does so in our higher realities. While by agreement, everything that exists consensually does so only within the specific courts, districts, and jurisdictions wherever this remains the case and only if you agree with them.

This is a model of the universe, and we are still defining here the main natural laws leading to our familiar natural laws of the universe, the ones based on mathematics, electrodynamics, relativity, and classical physics. We have just encountered Existence, and we have just defined universes and realities.

Existence defines this entire world objectively, from a perspective within this world. Existence has three natures, and they are relative to each another. Objective existence defines this objective world entirely, including its spacetime continuum, its field, which is the common gravitational, electric, magnetic, and electrostatic field, its radiation of all kind including visible light and the entire electromagnetic spectrum, along with all objects and living beings from this world, since these are objective in nature. Furthermore, Existence defines all objective life as seen from the perspective of this world, since Existence is the existence of Life, defining Life entirely, everywhere and continuously, in all her forms and realities.

Another nature of Existence is subjective, defining all subjective realities found within this world, as the multitude of computer realities and mind realities along with zillions of other inner realities held by our world. Therefore, all intelligences found within the multitude of mind realities are subjective in nature, they exist in a subjective manner, while as seen from the perspective of this world, they are alive in a subjective manner.

All existential natures are relative to each other according to your own system of reference, and therefore from the perspective of any reality, everything exists objectively there, as long as you are there. All intelligences are objective living beings from their own perspective within their own inner mind worlds. Similarly, everything is objective in nature wherever you dream at night, and this is why everything is credibly real in your dreams. You always consider yourself objective, because your own perspective is always glued on you. However, your soul considers you subjective, because you are in a lower reality, considering you in this manner just as subjective as all your intelligences, since from his own perspective, you are always in an inner world below him, and therefore you are always subjective. While for you, your soul and all higher beings are always highjective.

Highjective existence defines everything and everyone within our higher realities, as the multitude of higher beings as souls and all religious and spiritual characters that exist highjectively. You are subjective in nature as seen from the perspective of these higher beings, while your intelligences are also subjective in nature as seen from your perspective.

Now we can define existence itself. At the universal level of the wider world, we can refer to Existence as everything that exists. While within the wider world, at the level of the multitude of its realities and clusters of realities, we have to refer to each existence applied to each reality simultaneously. Yet in order to understand the individual existence of each individual reality, we have to study briefly realities individually.

There are countless of realities in the wider world. Each reality is defined as everything that exists objectively there, yet only from their perspective there, which it is similar to our world. You can never exit realities directly, and you cannot transfer objects, subjects, and information from one reality to another, but only personalized copies of information. This is the case because realities themselves contain everything that exists objectively, and they have their own continuum, specific only to them.

Let us see what realities are. I define realities using mathematical knowledge because mathematics alone can transcend realities, being able to define them with a similar statement. Because our natural laws of the universe are specific only to our world, and therefore I cannot use them to define all realities of the wider world. Mathematics alone seems neutral, correspondent, and therefore pertinent enough to offer a common definition for all realities. Yet you can use any neutral abstract reality to define them, as music, business, spirituality, or poetry, ending up defining realities in a specialized abstract manner. Yet even in this abstract manner, it remains an accurate definition of a reality, but only within that specific specialization. I use mathematics since it is present in the ultimate reality, and since I already have to use it through existence, because existence itself defines everything within realities using first level logic and therefore mathematics, among everything else. Coincidentally, mathematics is capable to define realities at all levels, covering them all, or most of them.

Our natural laws maintain our objective perspective, and this includes our spacetime continuum, our laws of classical physics, along with our constants and coefficients. Note that there is a distinction between the natural higher laws defining Life, Intelligence, and the wider world, and the natural laws of the universe defining only our universe, as classical mechanics, mathematics, relativity, and electrodynamics.

We state the definition of a reality right away, applying to all realities of the wider world, including this world. This should be at the base of the human reality, with everything in it, everything that you do, achieve, and experience throughout life, with you in it and everyone else. Yet since this is only a definition, it is only first level knowledge, still incompatible with Life, Existence, Intelligence, Interconnectivity, and the wider world, since these are of the tenth supreme level. Yet it is the best that we can do so, since again, these words and paragraphs can carry only a first level algorithmic communication. Furthermore, we integrate this definition of a reality within this entire mental model of the human reality, and this is of the third intelligent level. This is the case because this mental model of the human reality is not here in this book, but it is already in your mind, part of your third level intelligent conscious mind, and therefore part of your third level intelligent understanding of this world. While if you manage to integrate it accurately, lively, and in an intelligent manner, then it is of the third intelligent level. Let us see how you transform this first level algorithmic definition of a reality from the next paragraph, into a genuine third level intelligent conception of a reality in your mind. Yet by now, you had already conceived your third level intelligent conception of a reality, throughout this entire book. I present this definition as an enumeration of statements, helping you elaborate more.

A reality is a distinct set of objects and events spanning a commutative topological ring or a commutative topological vector space. A reality might also be a set of more than a continuum, objects, and events.

All realities are objectively real as observed from within, while nothing exists objectively outside them, not even other realities.

We cannot assume that other realities do not exist according with their own separate laws of existence.

Realities do not intersect objectively. They do not share information, components, and matrices of continuum in an objective manner.

You can transfer only interpreted copies of information between realities, and nothing else.

This was the definition of a reality, formed of these five statements. Right now, it might seem that this world is the only reality that exists, with the other ones only imaginary, as stated by science, but this is not the case. You can never define and model the human mind, the human reasoning, and life in general, if you do not consider the existence of other realities. Everything relates with the definition of Existence in the first place, and with the order or hierarchy of realities on another.

Study your computer hardware, to find digital fluctuations of electricity taking place in all its switches of the motherboard. These rapid fluctuations of electricity are digital. Electricity passes or does not pass through the switch, electricity is objectively real and therefore part of this world, yet the encoded fluctuations of electricity themselves are not part of this world. The electricity itself is objectively part of our world, the switches and the entire computer hardware are objectively part of our world, the light itself is objectively part of our world, yet the only subject not being part of our world is the intelligent coding itself made by the digital fluctuation taking place within the switches of your hardware, since that is subjective in nature, and not objective.

More precisely, the hardware of your computer exists objectively in our world, and as it is, it is capable to create a matrix of intelligent digital coding. This intelligently encoded matrix is capable to hold a different continuum from what we have here in our world. This new continuum is the digital platform or operating system of your computer, using computer language. This is a digital inner continuum, able to hold any software and any computer inner world, as the computer videogame "The Sims 4." This specific computer videogame is a different reality in itself, and it has a different continuum at its base, which is the computer operating system with its computer language, while it has different natural laws defining everything in this videogame, in any manner that they are defined by the creators of this videogame.

Note that as any created reality, you do not have to create your reality infinite, but you can only give the illusion that everything is infinite in the videogame, if you wish.

As stated, it is important to notice the difference between the naturally living, and the consensual or the agreed part from any reality. All natural realities, inner or higher, are part of Life, whenever living beings make them part of Life in any manner, but more importantly, whenever these new, inner made realities are part of these living beings themselves, since only in this manner, the inner realities can draw upon the living inner intelligences of the field. Natural inner realities are mostly related with cognitive systems, just as you form, hold, and maintain a zillion inner mind realities within your own cognitive system. Videogame realities are different, since they are based only on a digitally encoded matrix, therefore they are not part of the human cognition, and therefore they are not natural but artificial, and are not part of Life.

It is possible that the computer brain interface to become so advanced, to be able to link computers with the human cognition. In this manner, by using computer technology, you are able to connect with the living intelligences of the field, through the human cognition itself.

These are living inner realities, part of Life, because, by being part of your own cognition while you are alive, they are part of life and therefore they are natural. However, in the case of the computer brain interface, we still refer to the natural, living human mind as it is only aided by computer technology throughout cognition. Therefore, all related cognitive processes are naturally living, and they are part of Life, including the multitude of inner realities that you create in this manner, as this entire world.

Now we can understand better this world, since we can study if it is a natural living reality, or only an artificial nonliving reality. Do all living beings in the universe have their roots in the living intelligences of the field, and through the field, in Life herself? Is our universe or reality direct part of the higher beings, and through them, direct part of Life herself? Because this decides if you are alive yourself or you are only artificially created. Currently, all ideologies, old and new, including all scientific, religious, and spiritual ideologies, still consider you a genuine living being, with only your own consensual corporation that is your brand, trademark, or name written in uppercase letters considered consensual in nature. Through this corporation, you interact in society, in the Brotherhood, and in the Consensual Matrix, since only through your consensual corporation you are compatible with the

Consensual Matrix. Yet the universe, which is this entire world, is always considered to be natural, and therefore part of Life, even though this topic is never approached in science and by many religions. More precisely, there is nowhere stated the opposite, that this world is artificial.

Furthermore, if the universe is created and maintained using only technology, then it is not part of Life and of the wider world, but it is only part of a larger supercomputer. Yet with all these souls coming here, it becomes alive, and therefore it becomes part of Life. Furthermore, all higher beings cannot have a comprehensive emergence and experience in this world, here on Earth, if this world is only artificial. While through millions of testimonies, this world is part of a larger, common higher natural living mind, making it part of Life and of the wider world.

Even the Consensual Matrix itself considers human beings real, natural, alive, intelligent, and even of a very high status in many instances, divine. While continuously, the Consensual Matrix avoids living human beings altogether, since they are not consensual but of a higher status altogether, while it always makes sure that it has all the necessary permits, certificates and licenses to operate in this world, here on Earth, not to get in trouble. Officially and legally, it is not you the living human being serving the Consensual Matrix, but it is the other way around, the Consensual Matrix always serves you. Yet the Consensual Matrix makes sure beforehand that you are only your consensual corporation when it deals with you. In this manner, the Consensual Matrix can do anything it pleases with you as a corporation, since the Consensual Matrix owns your corporation entirely, whoever you are and wherever you are positioned in society.

More precisely, if you manage to state your living intelligent human nature, the entire Consensual Matrix along with all politicians and all public servants serving it must serve you unconditionally, since as an intelligent living human being, you are significantly more superior and more important than the Consensual Matrix and all its jurisdictions and corporations.

The primary meaning of the Consensual Matrix is to serve the intelligent life throughout the wider world, since this is why it is instated and it is maintained in the wider world. However, as long as the Consensual Matrix has all the necessary legal means to consider you a corporation, then it is the other way around, you must serve the Consensual Matrix in all details, because as a corporation within its jurisdictions, the Consensual Matrix owns you, and can do to you anything that it wants.

The Consensual Matrix always considers and always addresses only the consensual part of the universe, and even then, the consensual part of the universe takes place through you, through your own direct or implicit agreement to be consensual yourself, whenever you choose directly or implicitly to identify yourself with the multitude of your documents, since they have on them your name written in uppercase letters, which is your consensual trademark or corporation.

Are we capable, from here, from within our world, to tell if our entire world is ultimate or naturally made, or it is only created, naturally or artificially? We have a multitude of inner, created, and naturally made types of worlds to use as a reference, as all computer inner realities and mind inner realities, yet even so, this consideration is only empiric, objective in nature, and therefore it can be accurate or not. Furthermore, our answer must relate accurately to the natural laws of our higher realities, and not only to the natural laws of this world, since Life defines all of them, but only as long as they are part of her.

The answer might be as simple as observing yourself, yet this answer is empiric in nature, since all observations are empiric. You certainly can feel the life and natural consciousness and intelligence in you, through all your natural needs and meanings, and therefore you know that you are alive. Yet all these natural needs and feelings can be given to you directly, by default, through higher laws, to make you only feel that you are alive, while you could be only artificially created, similar to all book characters and videogame characters. Your daydream characters are alive, since they are part of yourself, if you happen to be alive. You might have been created yourself artificially by higher beings, to give you the impression that you are natural and alive, while you could be only artificially created.

However, with the Consensual Matrix always around and always exploiting you, while always considering you a corporation, it is more likely that you are an intelligent living being part of Life and the wider world, while the Consensual Matrix tricks you continuously to remain only a corporation. Because if you were created artificially, in an entire artificial world, then our Creator placed at the base of this world all consensual laws meant for servitude directly, while in this manner, forcing you directly, by default to serve, similar to all artificial intelligences of all computers, also serving continuously by default.

This distinguishes between the two types of religion and spirituality encountered throughout the book, the naturally living ones and the consensual ones, since many significant religions speculate exactly on this idea, that you are actually created artificially to serve, while you are only assuming that you are alive. Which is not true, yet if you consider yourself only an artificial or consensual creation, you must serve from then on continuously, which is exactly what the Consensual Matrix wants.

For the Consensual Matrix, it does not matter if you serve through religions, political parties, jurisdictions, consensual Brotherhoods as the entire masonry, cults, places of employment, dictatorships, criminal cartels, or military units, because as long as you serve, you serve consensually, and you serve the Consensual Matrix, which spans most of the wider world, either as a soul or human being, and the Consensual Matrix is successful exploiting you. Yet if the Consensual Matrix cannot reach you, if you happen to be more developed, then you escape its servitude, and it cannot exploit you, it cannot profit. Notice how the Consensual Matrix cannot exist anymore where you are, if all people manage to escape it, and it has to move away to operate somewhere else.

You are alive if you have a higher self or soul, and if this is alive. Yet how exactly can you tell if this is the case or not with you, mostly when ideologies teach you both possibilities, that you do or do not have a soul, depending on ideology? There is one way to tell if you have a soul, since if you experience higher events in your life that cannot be explained in any manner, as higher forms of events, higher circumstances, or higher cognition, the kind of interesting paranormal activities as slight as these can be, then you are genuinely alive yourself, you interconnect with living beings yourself, and furthermore, this states that this world is genuinely natural, real, and alive itself. Yet as always stated, this is still empiric in nature, and therefore you cannot consider it fully, as long as you cannot relate with the natural laws of the higher worlds, or directly to the natural laws of Life and of the wider world.

There is yet another way to tell if you have a soul, through the level of your needs and feelings. Life starts with the second developmental level, which is the animal, intuitive level. The zero level is the addicted level, the first level is the ideological, servitude, consensual level, the second level is the animal level, while the third level is the intelligent human level. If the feelings, needs, and meanings that you have and fulfill throughout life are at the first two nonliving levels, the zero addicted level and the first consensual, hierarchic, ideological level, then you might not have a soul, and you are only a living human being. While if you are able to identify and fulfill your second animal needs and meanings along with your intelligent human needs and meanings, then you live your life at your own intelligent human level, you are alive, and you might have a soul.

As a reference, if you are capable to follow this book, if you are capable to reason alongside this book, while forming a genuine intelligent model of the human reality alongside this book, while being rewarded continuously with good feelings for your continuous learning and development, it means that you have a soul. Furthermore, if you feel your higher self enjoying specific knowledge in this book, as knowledge about the universe and about other realities, or about intelligences, society, interconnectivity, computer realities, reasoning, inner self, or feelings, then this is your soul experiencing this world with you, and this is actually how it feels.

As another example, there are millions of testimonies coming from those who had left this world and then came back to tell their story, and everything that they state remains consistent with a natural living world, part of a wider, larger cognitive system. This world is alive, and it has a cognitive nature, being direct part of Life, Intelligence, Mother Nature, or Universal Mind, call it as you please.

Even when the entire Consensual Matrix considers this world, it never states that it is not natural or not alive, but it only ignores it or it only refers to the consensual part of the world, which manifests only through your own consensual agreement, but not through a possible artificial nature of the universe. The Consensual Matrix never states that the human reality is natural and therefore alive, while it never states the contrary either, that the human reality is consensual in nature.

In general, the Consensual Matrix is just, very precise, and very consistent in everything that it states and does. While every time it does anything against Life, since it happens often, it does so only through you and through your own direct or implicit agreement. Therefore, you are always the one assuming your direct or implicit responsibility for everything that you do for or through the Consensual Matrix, but not the Consensual Matrix.

As another reference, society is full of documents, permits, certificates, stamps, requests, and signatures, because this world is not consensual in nature by default to make all consensual agreements valid by default, since they are never compatible by default with this world. The entire consensual of this world must always be made valid through you, through your own consensual agreement that you allow them to apply to this world and to you, because you, humanity, and this entire world are neither artificial nor consensual by nature and by default. Therefore, regardless of what you are made to believe by all ideologies and jurisdictions, you are the one allowing the entire Consensual Matrix to interact with this world, through you.

For example, it is you, the living human being allowing all corporations to manifest here in this world, through your direct permission as a living being. More precisely, there are specific authorities, the presidents, acting as proxies of all living human beings of each nation, and these authorities give the necessary signatures allowing all corporations and jurisdictions to exist and to operated here on Earth, in order to do business, exploit, and profit here on Earth. You allow these authorities to be your proxies every time you vote and every time you agree directly or implicitly that they can act on your behalf. This is the case because this world belongs to living human beings, at least the surface of Earth, and therefore the living human beings can choose everything regarding this world.

This is a living, natural, real, intelligent world, part of Life and of the wider world. While it is more likely that you have a soul, since there are not too many available spots left in this world, all being taken by the billions and more, with many souls coming from all higher realities above, since in this entire cluster of created realities, all or most of the souls have souls.

It is different within the natural realities of Life and the wider world, because intelligent living beings have classconscious intelligences throughout the natural realities of Life and the wider world, called shems or spirit worlds, which are different than highconscious intelligences with all souls included. Throughout intelligent golden human ages, humanity also has its classconscious intelligence or social spirit world functioning well, allowing all humans to remain interconnected naturally and intelligently with everybody else in a common intelligent human mind, where you can fulfill all intelligent human needs and meanings together in an intelligent manner. Telepaths achieve this intelligent classconscious common mind with ease, yet throughout dark ages, humanity is never allowed to form it, because it makes tyranny, servitude, and therefore exploitation impossible.

Therefore, in our cluster of created realities, humans should have four main intelligences, the conscious, subconscious, classconscious, and highconscious intelligence. While throughout the natural worlds of Life and of the wider world, living beings have only three main intelligences, the conscious, subconscious, and classconscious intelligences.

It is also important to understand the entire relationship between living beings, the reality that they inhabit, their specific worlds, societies, and interconnectivities that they form, the specific natural and consensual laws governing their life, fulfillment, and achievement, and the manner in which these laws originate, along with their nature and relationship with Life and with the Consensual Matrix. As you notice, it takes over fifty books of this series to model all these additional topics of study, because the human reality remains within this world, and therefore it is only objective and consensual in nature, while we have to use knowledge of the human mind, existence, meaning, higher reality, life, development, reasoning, society, fulfillment, needs, and feelings.

All realities are made or created naturally, artificially, fictionally, consensually, or digitally, from the outside, from a higher reality, and this is how they remain inner realities as seen from an upper perspective. There are other inner realities created in our world, and they count in zillions. These are our mind realities, and they are formed in the exact manner, through electromagnetic fluctuations formed within cells, neurons, and cellular components. All brains form inner realities while reasoning, and not only computer hardware, yet they use similar intelligent fluctuations of bioelectric currents taking place everywhere throughout the body at subcellular levels, creating the conscious, subconscious, and highconscious intelligences as these live within all their inner realities by the zillions. Your daydreams, regular dreams, and regular planning are examples of inner realities or mind realities, and I model these throughout dozens of books of this series "Human."

The human brain and body hold the human mind in a

subjective matrix, formed by the encoded electromagnetic fluctuations within cells, brain, and body, along with similar ionic fluctuations, patterns of charge distribution fluctuations within the cellular membrane, along with specific ionic, amino acid and protein oscillatory movement and therefore electromagnetic fluctuations in the electromagnetic field, while this is the basic electromagnetic type of life.

The human mind is not an inner reality in itself, but it is the summation of all mind inner realities formed, maintained, and used throughout thinking and reasoning. These inner mind realities are capable to hold a multitude of intelligences, while these specific intelligences perform all human cognitive activities, including the human thinking and the human reasoning, with one specific intelligence from among all these being you, the conscious intelligence.

One specific mind inner reality worth studying is your conscious intelligent inner replica of the world, which is your actual intelligent conscious mind reality, consisting of all memories and understandings that you have of the outside world, gathered throughout life through your continuous learning or experience. You are your intelligent inner self within this replica of the world, and here is where you undergo your intelligent reasoning consciously. Your intelligent conscious mind is an inner replica of the outside world, because throughout life and throughout learning, you manage to place in it all memories, all knowledge, and all understandings of the outside world, while in this manner, you form an entire intelligent inner replica of the outside world, spanning the cortex. The cortex itself had been developed by the organic form of life in order to offer you intelligence, while offering you the possibility to understand everything in a conceptual intelligent manner, by using this entire intelligent inner replica of the world spanning the cortex.

There are similar subconscious inner replicas of the world in your mind, specialized on all domains, all created by your specialized intelligences, as your eating intelligence, reproductive intelligence, or recovery intelligence. Everywhere within all inner realities of your mind, there are countless of intelligences and inner intelligences living life normally, since everything within all realities exists objectively from your own perspective, for as long as you are there.

We are already observe the relative nature of existence in this world, in the human mind, in our cluster of created realities, and in the entire wider world, objective, subjective, and highjective. Therefore, everything found within any reality exists objectively, and these realities span as far as they exist objectively, as far as their own continuums spread. All inner realities formed and held by these realities are subjective in nature, and they can be of any kind and of any form, besides the two cases taken above, the mind realities and the computer realities. Simultaneously, from the perspective of all inner realities existing subjectively, the outer reality forming them, creating them, and holding them in their matrix is highjective in nature, and therefore everything existing above them is highjective in nature. This is the case with all souls and higher intelligences including all higher intelligences calling themselves deities and seeking veneration from humans throughout the dark ages of Earth, all claiming to be the Divine himself.

The Divine himself is also a higher living intelligence, the supreme living intelligence, including everything that exists highjectively, objectively, and subjectively, everything that exists everywhere and continuously, since the Deity is omnipresent, omniscient, and omnipotent.

Mario from the videogame is subjective in nature, as it shows on the display of your laptop. Mario appears in this manner on all laptops, subjective. Yet there is always the replica of Mario in your mind, and this is subjective just as well, because in your mind, you also have a Mario world, directly correspondent to the Mario videogame world from computers, yet this is alive, not digital. It is cognitive subjective, and it is alive. There are different Marios to consider, with those from our minds enacted by our genuine living inner intelligences, the way actors play their roles.

It is through similar acting intelligences that you always

reason alongside the conscious intelligence, and you reason in this manner not only during videogames. These similar living acting subjective intelligences from your mind are doing all your reasoning through mental simulations or mental models.

Therefore, Mario is subjective in nature, while you are highjective in nature from its own perspective, if it can ever reason to study you, its player. Souls are just the same, they are highjective. They are not yours, but you are theirs, you are their subjective avatar. The word 'avatar' is thousands of years old, and this explains the entire subjective existential nature of this entire world, from the perspective of all higher beings to have ever roamed this world through all their avatars. This is hidden by science currently, while these specific paragraphs are erased and censored from all bibles and records. This is why when you find texts as: 'let us get down into them' or 'and they found men or daughters of men worthy,' they refer to avatars and to a forbidden word: incarnation. Reincarnation means that you come back. Why not coming back here? To go elsewhere, wherever they take you, forever. Because what you find in science are truncated empirical models censoring away from you the rest of realities of the wider world, along with the wider world itself.

Are there two parallel realities or universes in the world? If they communicate somehow, if they exchange matter, continuum, or information in any way, then there is only one universe, which includes both sides. If you were capable to open a wormhole from one universe to another and went through, then there are not two parallel universes anymore, but a single one. If there are two parallel universes, but you cannot manage to open a wormhole ever, then how do you know that there is another parallel universe out there, since nothing exists outside our universe? Furthermore, the notion 'out there' is impossible to use. If you have managed to open a wormhole only temporarily, then it is still a single universe containing two separate sides, which had come in contact only temporarily. Now the pathway is closed, and there is only one universe left, yours, with nothing existing objectively outside.

The Human Reality

As you notice, it never matters in which universe you happen to find yourself, because that is your universe, your objective world, which is your objective reality, and nothing else. Throughout your dreams, you exist in each one of your dreams objectively, for as long as you are there, with this world subjective or highjective in nature, depending on where you dream your dreams. Because there are astral planes that you can access throughout your dreams that are higher in existential nature than this world, and therefore everything around here is subjective in nature compared with you up there. These are the cases when you dream with your soul, or when you remember your dreams from your soul's casual higher experience.

If you play your 'GTA5' videogame for hours in a row, you lose connection with this world, and you exist in the videogame entirely. Technically, you exist there objectively, while in those specific minutes and hours, the outside world is highjective in nature for you. Because many times, you are not the entire physical body and cognitive system, but you are only the conscious intelligence.

Note that you are capable to access other realities through you, through your higher and inner selves, and nothing else. You can also access other realities, higher and lower, through all other living beings and intelligences through which you manage to connect. Yet outside yourself and outside your interconnections, you cannot access other realities, you cannot even consider them higher or lower since the laws of existence do not even apply to them, and therefore they do not exist for you, regardless of how much you try to imagine or understand them. This is the case with the continuously interconnected life forming Life or the wider world, since the Life or the wider world are everything that exists everywhere ever, nothing more and nothing else, because Existence defines Life or the wider world just as far as they exists. If there is anything else past Life and the wider world, then that is also part of Life and the wider world, since it exists.

Let us test the contrary of our model of the universe right

away, to see how our model holds consistency. What if we define the universe to be non-unique, allowing the possibility of having multiple universes to exist objectively, even if this definition is not consistent with the real definition accepted by all sources, present and past, that the universe is a 'uni-verse,' or one verse only? If two universes exist objectively simultaneously, then they are held by the same matrix within the same higher realities, and therefore the two universes are one, united by the same continuum. Whatever you do, you cannot have two separate universes held by the same matrix, holding the same continuum, and therefore allowing them to exist objectively simultaneously in parallel, since that is only one universe, made by both realities simultaneously.

For example, for you to go from one reality to another, you have to go from one entire medium to another, mediums holding different matrices and therefore different continuums. You can go into someone else's mind, which is more common, or you can go through your soul to other related realities, while you can always tell when this is the case, if the feelings and senses of perception are more or less intense there.

Where exactly are you throughout your dreams? Science defines dreams as pure random aberrations. However, you might end up in any reality throughout your dreams, since these switch constantly. Just study the laws of physics throughout your dreams, and if you find them similar to the natural laws of the universe, then you are within lower realities. Yet if they are entirely different, and more importantly, if they are still consistent among themselves regardless of how different they are, then know that you are within important higher realities, so make sure that you behave properly, because it counts.

We know about our world that it includes all stars and planets, all galaxies, clusters and super clusters, with nothing left outside. The universe includes all matter, everywhere, every time. The universe also includes all light, all electromagnetic radiation. Energy defines both matter and electromagnetic radiation, so the universe includes these both, and more importantly, it includes the substance that both waves and matter are made of, and this is the electric, magnetic, static, electrostatic, electromagnetic, and gravitational, which is the field. Consequently, the universe includes all field, being it in the form of matter or electromagnetic radiation.

From the official definition of the universe, we find that the universe includes all space and time, which is the spacetime continuum, holding directly the field. From the etymology of the word 'universe,' we find that the universe includes all events, all movement, and all information. The universe also includes all the natural laws making it possible. I expand these to include all laws in the universe, as accurate science, psychology, and history, since they are based on the natural laws of the universe for as long as they remain accurate, and for as long as they interconnect with each other. Usually, only beliefs cannot connect with themselves and with the accurate facts, because accurate facts always connect with accurate facts, furthermore connecting with the natural laws of the universe, which are our accurate natural laws here in our world.

Note that these laws are not only physical or cosmological in nature, since the universe does not hold only substance and spacetime, but it holds art, life, religion, and society, forming the very complex structure of the universe. From the definition and characteristics of Existence and realities, you define reasoning, intelligence, and life. In this manner, you define biology, psychology, and society. From here, you define sociology and history. While defining everything, you manage to link everything to the natural laws of this universe, and to the natural laws of the wider world above.

I refer to our universe as our world, our reality, or the real world, while the wider world is Life, Intelligence, or the One, everything that exists everywhere and in all realities. By being 'everything,' it implies that the universe is one, unique, which further implies that it contains all its components mentioned above, which further implies that nothing exists outside the universe, not matter, neither electromagnetic waves, nor space, time, or information, but they are inside the universe. The wider world is Life or Intelligence, it is unique, it includes everything, and it is everything that exists, including all life, intelligence, consciousness, and interconnectivity.

What if there are parallel universes 'outside' our universe? We can still presume the existence of other universes outside, since anything can exist outside the universe in anyone's imagination. However, nothing can exist objectively outside the universe, otherwise it is part of the universe. There is nothing defining the notion 'outside,' since, by definition, everything real, everything that exists is inside the universe. Those universes that are outside, once they exchange field, spacetime, or information, allowing you to comprehend their presence and existence, become part of our universe by definition, since that existence defines our universe, since our universe is everything that exists objectively. If the universes found 'outside' do not exchange field, spacetime, and information with us, then we can never perceive their presence, which further implies that they do not exist objectively for us, they do not exist at all.

What if we can detect another universe 'out there?' Then through detection, the information coming from that universe makes it part of our universe, since our universe includes all information, which implies that the newly detected universe is part of our universe, or we are part of that universe. In any case, there is only one universe.

We are consistent with official sources, since they state the same definition, that the universe is everything that exists. We only add the word objective: the universe is everything that exists objectively. Science never exits this objective world throughout its research, and neither does psychology. We also consider etymology.

Why is etymology important? Etymology relates directly with the laws of the universe, since it consists of terms and concepts, remaining accurate knowledge by default, regardless if words themselves cannot transcend from one reality to another. Yet since their concepts are unique and since concepts can transcend realities if it is the case, then etymology relates directly with the natural laws of this world, as defined by our higher reality. While words can be anything that you choose to invent as a word for each concept.

Because the meanings of many words are changed throughout time to define something else, or to define even other words altogether, while the original meanings might be lost or banned, yet their concepts cannot be changed, since they are unique and therefore accurate. In this manner, you can still interpret the original concept associated to each word in use, old or new, through etymology alone.

As an example, the word 'demon' is a very bad word currently, while in the past, it meant the highest demi-deity, and everybody prayed to demon or to de-Amon. While you do so even currently, whenever you exclaim Amon, Amen, or Amin. Demon comes from the 'dem' which means semi, and 'Amon' is 'on' or 'an,' or 'An,' the supreme Sumerian deity. All the above are not transitions, but they are various forms of the name given to this specific deity throughout various nations, including Sumer and Egypt. This is also related to the name 'Amen' that you can use throughout your prayers, since these are two family related deities. Implicitly, the word 'demon' or 'demi Deity' means the offspring of the Deity and men. The demons were the deities of Sumer that we have just studied and we could never decide whether they were spirits, human beings, or spirits of human beings. The three can also be correct at once, in which case, they are still us, and they will probably be with us until the Day of Judgment, as stated. This word did not reach us directly from Sumer, but it passed through Egypt.

Demons were later on associated with Paganism and Arianism to define both good and bad spirits, or angels, good and bad, embodied or not. Afterwards, the Coptic religion came and kept some of these names, it started Christianity, and then, when Christianity came, it banished all old beliefs and values, good and bad alike, along with all demons, good and bad, to keep an only deity, the Deity. I always refer to the unique deity as the Deity or the Divine, to maintain neutrality.

We see how everything is related in the wider world. Furthermore, everything happens for a reason, because the wider world is everything, while all events are interconnected throughout all lines of existence defined by the Existence of Life.

What is the relationship between this world and our higher reality creating, holding, and maintaining this world in its matrix? One is higher and the other one is lower in existential level, but we are interested in its meaning, since it has a cognitive meaning, as stated previously. Spirituality states the same, but how can it be? How does everything integrate in Life and in the wider world? How do you integrate in all these as a living human being? How does society integrate in all these? You have to understand the human reasoning in order to understand the integration of this world in the wider world, through its cognitive nature. How do you reason? Let us see.

You reason through mental models, and you do so within your own inner replica of the world, which is your actual intelligent conscious inner mind world spanning the cortex. Why reasoning? You need to solve your problems that you encounter in the outside world while fulfilling your needs and meanings, and you have to figure out a way to do so in your mind first, before you apply everything in the outside world. This is how you model everything within your mind first, while if you fail in your mind, you do everything all over again. You persevere, and you do everything faster, in your mind first, even repeatedly if you fail repeatedly, since it is safer, faster, and easier to mental model in your mind, than to do everything directly and unprepared in the outside world. While everything that you do in the outside world, your entire human activity, you do in order to fulfill your needs. It is important to be successful in your mind first. It is important to find your successful ideas in your mind first, you apply them in the outside world through your physical activity, you fulfill your needs, and your problem is solved. Yet in general, the problem is not solved, because the toilet still leaks, it exasperates you, mostly since you cannot afford a plumber, while this is basic life in a basic consensual world.

How do you reason intelligently? You reason intelligently only within your intelligent mind world spanning the cortex since it is impossible otherwise, because only the cortex offers intelligent reasoning. While you always employ throughout your intelligent reasoning all intelligent memories that you acquire throughout life, whatever necessary, while avoiding the rest, since you have to use specific background intelligent knowledge throughout the entire intelligent reasoning and throughout all intelligent mental models, towards finding your successful solution, which is also intelligent knowledge, intelligent information, or intelligent memories.

You do not form your new intelligent knowledge or idea out of nothing in your mind, but you conceive it there after some time and effort, in a living manner, since you give birth to your intelligent knowledge, intelligent solution, or intelligent idea in your intelligent mind spanning the cortex, because all intelligences are alive.

When you learn anything from the outside world, you simply conceive its replica or counterpart in your intelligent inner replica of the world, through everything that you already know intelligently in that subject, since this is how all intelligent elaboration takes place. It is similar when you conceive your intelligent ideas or intelligent solutions towards the fulfillment of your needs, since you also conceive them in form of living intelligent conceptions, by placing together all the necessary previous knowledge in that subject in form of a very specific systems of intelligences, and this is how you conceive all your intelligent knowledge throughout your intelligent inner replica of the world.

You seek to find this successful knowledge, because you need it in order to fulfill your needs. Yet many times, you cannot fulfill your needs, because you do not know how. This is why you have to reason intelligently, in order to find this specific needed knowledge, your successful solution, which is your successful idea. While you do not have it, because it is missing, obviously, and you do not know it yet. It is missing either from you general background of intelligent knowledge spanning the cortex, or it is not available in the outside world altogether, and you must figure it out on your own, which is very tedious.

There are many types of thinking, with the most advanced form of thinking as the human intelligent reasoning, which includes mental modeling, cognitive digestion, elaboration, intuitive thinking, and algorithmic thinking. It is important to know more about your memories and the manner in which you have, store, elaborate, and use them, since you have an entire inner cognitive world holding your memories in your mind, and it is important to know it.

As a reference, right now as you read this book, we have to find on our own all knowledge related to the human reality, since it is unavailable in the outside world. While this entire model of the human reality that we form in this book is actually our comprehensive intelligent mental model necessary to find this entire knowledge. It is very complex, since reality and humanity are very complex, yet it is still a third level intelligent conception in your mind, as complex as it is, because this book goes through all the necessary details to make it possible.

Successful solutions are successful ideas, while these ideas are simply missing information that we work hard throughout reasoning to find ourselves. It is tedious to find successful ideas, while once found, these successful ideas can change our lives for the better in many instances. This happens often throughout life, since everything is part of the human knowledge, an intelligent human achievement for the humankind. While in order to find intelligent knowledge ourselves, we have to reason.

You create intelligent mental models throughout your intelligent reasoning, which are perfect simulations done within your inner replica of the world, yet they are correspondingly similar with everything happening in the outside world. All memories and understandings that you manage to gather throughout life are not simply dumped in your mind, and they are not simply found in perfect libraries of data, but these memories are placed exactly as their real counterparts are found in the outside world, for an increased consistency with the outside world. Therefore, your inner replica of the world is as similar with the outside world as you manage to create it throughout your learning and life experience.

You as a conscious intelligence are the creator of your own inner replica of the world holding all your conscious memories, while all intelligences of your cognitive system form their own inner replicas of the world, through similar learning experience, with many of your intelligences having direct access to your senses of perception and your own conscious inner replica of the world. You replicate yourself in your inner replica of the world from who you are in the outside world, as your inner self. You actually live your life as your inner self within your own replica of the world, while you are capable to roam around freely within your replica of the world throughout your thinking, reasoning, feeling, daydreaming, mental modeling, researching, and learning. In this manner, you can retrieve memories, you can experience memories, and you can daydream and plan.

Note how you do everything objectively in your inner world as an inner self, casually, as though you live a normal life within your replica of the world, since all realities are objective as long as you are there, experiencing everything objectively and materially firsthand. This is how reasoning and mental modeling is done, always casually and objectively within your mind by your inner self, by your inner characters, and by all your inner intelligences, as though you live your life normally. Everything exists objectively in the reality were you happen to be, and therefore everything is objectively real there for your inner self and for your inner characters or inner intelligences. Because you also have in there the replicas of the real people from the outside world, and alongside them, you as the inner self, experience everything that you wish to experience within your inner replica of the world. This is how you daydream, tell stories, invent, write books, read books, and mental model all your social circumstances. This is how you perform your mental models, and this is how you perform your reasoning, planning, and daydreaming.

There is more to your memories and to your inner replica of the world, since all memories are intelligences in themselves, with the memories or replicas of real people including yourself as genuine living intelligences, living and roaming freely within your own replica of the world. Your other intelligences have access to your inner replica of the world, and it helps them create their own specialized replicas of the world. They think and reason there, they send you needs from there and you have to fulfill them, you have to reason before you do so in order to figure out how to do everything, while your subconscious intelligences help you continuously, by modeling, acting, and simulating everything alongside you. They do everything by themselves and then they only pop up the final idea in your mind, and you use it as it comes. This is how you reason, this is how you behave in the outside world, this is how you succeed, and this is how you live your life.

It is important to see how you use your memories or acquired knowledge throughout your reasoning, since they are part of your mental models, using your own previous knowledge and understanding of the circumstance that you mental model in your mind towards finding a successful solution. The more accurate your previous knowledge is, the more accurate is your reasoning, and therefore the faster you find your solution or successful idea, while this makes the difference between not being able to get by since you cannot do anything on your own because you do not know how, and finding the most ingenious ideas to fulfill your needs.

Therefore, the more accurate your knowledge about the outside world is, the better your reasoning is, and the more successful your ideas are. However, over ninety-nine percent of your knowledge is based on beliefs, stereotypes, strong personal convictions, and entire ideologies. This is consensual knowledge that is not necessarily accurate, yet you undergo an entire consensual existence through them while always assuming that they are accurate, and you cannot succeed in the outside world without the help of the Consensual Matrix, because you cannot reason on your own and therefore you cannot find successful solutions on your own. Ninety-nine percent consensual knowledge is only an estimate, but if you consider entire erroneous models of the world as the big bang theory, which always stand at the base of your inner replica of the world while compromising your entire intelligent reasoning, it ruins your entire cognition, mostly while trying to match your human meanings with the meanings of Life and of the wider world.

You are your conscious intelligence within your cognitive system and you reside in the left prefrontal cortex as the intelligent inner self, because through dendrites and axons, you manage to connect directly with your inner self while becoming entirely your inner self, which is the memory of yourself from your inner replica of the world spanning the cortex.

This is how you live your life in your inner replica of the word as your inner self every time you must reason, think, feel, and mental model, while looking, feeling, and behaving exactly as you do in the outside world. Because you never reason with your brain, with your mind, with your conscious intelligence, or with your intelligent inner self, but you reason **as** the conscious intelligence or **as** the intelligent inner self.

You walk around in the outside world as the physical body, you reason intelligently as the intelligent inner self, and you interact with your higher world as a soul. You are mind, body, and soul. Notice how you do all these in a separate reality as a different self, while you cannot mix them. You cannot reason with your brain, because the brain is objective and material and exists in the outside real world, not in the human mind. You do not even reason with your inner self, because your inner self is not part of yourself, but your inner self is you, and therefore you reason **as** your intelligent inner self.

How accurate are you throughout reasoning? You are just as accurate as you manage to learn and transfer information from the outside world to your inner replica of the world throughout learning, and then back in the outside world throughout reasoning and through your entire activity in the outside world while you fulfill your needs. As a reference, just take a pencil to create a drawing of your surroundings, and this is how perfect your replica of the world is. This is your one percent accuracy, with beliefs, stereotypes, addictions, underdevelopment, theories, and entire ideologies included, while this is the case with everybody. You will always find everybody arguing about outside objects, subjects, and circumstances, while these might be always fine outside, but they are inaccurately understood within their own replicas of the world. The more you learn, understand, and develop, the more accurate and pertinent your replica of the world is, along with your reasoning and successful ideas. Furthermore, the more you are capable to cooperate and maintain harmony among all intelligences of your cognitive system, the more they help you, the more successful in life you become, and the more you develop.

Many times, you do not reason consciously throughout your intelligent reasoning or throughout mental modeling, but the replicas of those around do so within your replica of the world along with all your intelligent conceptions of everything that you know and understand, mostly when you are trying to solve social problems, because the replicas of people from your inner world act and behave exactly as their counterparts from the outside world. This is how you predict the people around, this is how you always manage to predict future social circumstances, and this is how you always remain prepared for everything that can ever happen, because the inner replicas of your acquaintances and loved ones have already behaved that way in your inner replica of the world hours in advance, they have already informed you of what happened and of what they did in order to solve the problem, and now you know exactly how to behave in society in order to avoid that specific problem.

This is how you behave socially if you remain aware of how

you reason. It is also possible that your inner replicas of people remain connected with their outside counterparts, more or less significantly, deepening on your higher cognitive abilities. When the outside counterparts happen to depart, you still remain connected. While sometimes you do not know it, and so you ignore it.

Do you see how there is a resemblance, a continuation, or a correspondence between the outside world and your inner replica of the world, between your cognitive behavior and your physical and social behavior, and between society and your inner social characters from your inner replica of the world? There is also the correspondence between this world and our higher world, while this is the supreme law of correspondence, since it is found everywhere throughout Life, being a main supreme characteristic of Life.

Just the way you have created your inner replica of the world as a conscious intelligence, all intelligences create their own replicas of their world in order to allow them to reason, since they reason similarly within their own specialized inner replicas of their worlds, while they are conscious, and this is how they are capable to perform all their inner specialized tasks throughout the organism flawlessly. Furthermore, just as you create your replica of the world and reason within, in a similar manner, the creator or creators of our world reason within this entire world, and they reason within it through you and through all those around, through your behavior, and through all their behavior. While you are living your lives casually, since you are simulating or modeling higher events and circumstances in this manner. While everything that you experience here in an objective manner is a simple higher reasoning, an intelligent reasoning, or a collective daydream, whatever it is, it has a cognitive meaning throughout the higher worlds, with you always caught right in the middle of all these.

Who exactly is the inner higher self from among everybody? This is the question, since this is the avatar of our Creator. Many nations and religions could keep track and distinguish the avatar of our Creator throughout time, the incarnation of our Creator, later on to become the messiah or the son of our Creator, depending on religion or school of thought. He is here currently, and many people claim that they know him. Blessed be our Creator.

Souls or higher selves, which are the higher counterparts from our higher realities, happen to have higher abilities, including direct manifestation and telepathy. You also have these abilities throughout your dreams and projections, but not here in this world, yet it depends. As a telepath, you are capable to create entire inner replicas of the world alongside others, or you can enter other people's inner replicas of the world or a common replica of the world, to reason or daydream alongside them. This is what you do as a soul, you reason, mental model, or daydream alongside others, alongside your loved ones, even in this world. You decide together what specific reality or common reality to enter and experience together, while these can be extraordinary sometimes, having millions or billions of souls experiencing the same existence simultaneously within one same reality, as it is the case with this world. These commonly created realities can be smaller or larger, better or worse, safer or more dangerous, more or less dense, more or less interesting, and more or less educational.

This is the nature of this world, while society is the stage, the act, the mental simulation, mental model, or the daydream of the souls, taking place through you the human beings, the main characters or avatars of this world. Therefore, hold on to your loved ones, because you might be soul mates up there, and this is how you live your lives, one lifetime after another. Make this life meaningful together, since it always counts.

Now you understand how our world can be as vast as the entire universe in principle, since it is made in the replica of the higher world that can be as vast as the entire universe also principle, while you never know. Because all created realities are not replicated entirely, or they are only a work in progress, which is mostly the case. Realities do not have to be replicated entirely, since whatever has to be simulated or daydreamed can limit only to Earth, if this is what the souls desire up there, with the rest of the galaxy not actually here, but only lights in the sky. The Moon does not have to be there objectively, since nobody goes there. All tides can happen everywhere by default, even without a material moon, because they are programed by default, to happen with or without the Moon. It is similar with the artificial satellites of Earth, because they do not have to be in the orbit in order to transmit information, since all information can be transmitted by default.

Therefore, this entire mental model of the human reality can apply to our world only as far as it spans, while it always applies to the higher world as far as the higher world spans, even covering the entire universe, if this is the case in our higher world. Yet with our higher world made in the correspondent image of its own higher world above, our mental model of the human reality holds for them even better, because hopefully they have a full galaxy up there, or at least a full solar system with the Moon intact. If not, then it should be the case several higher worlds above, where our Creator lives, where they actually have a universe as seen in the sky, with all spaceships and aliens intact. Yet we already know the answer, because all mental models are made only in truncated sceneries that include everything necessary, with the rest removed systematically through focusing or concentration in order not to distract you.

Our mental model for the human reality holds for all higher realities above that maintain correspondence with this world. This is the case not coincidentally, but this is how all mental models are created throughout all lower cognitive worlds counting in large numbers, repeatedly and very fast, until a successful solution is found, on behalf of all higher worlds above, up to Life herself if necessary. This is the natural law of Mentalism, a supreme characteristic of Life.

This is how the knowledge and the reasoning that we generate through our model of the human reality is useful up there through us, because your entire reasoning, behavior, decisions, and achievements that you undergo, encounter, and manifest in society and in this world, you do on behalf of those above, on behalf of our higher realities, and on behalf of our Creator himself and above.

Everything that you do here is part of his cognitive activity and it is done for the worlds above, all the way up to the Devine. You do everything for him, and this is how you fulfill the Divine throughout life. You do so through your life, through your normal, natural behavior and experience in your life, through your reasoning and feelings, and you do everything through your decisions, needs, and expectations, everything defining you in life, so make it a good one. While you must always keep everything accurate, pertinent, and consistent, because otherwise, through harmful consensual needs and meanings, you end up against our Creator and against Life and the wider world, and it happens often, mostly throughout dark ages and consensual worlds.

Let us further test our developing model of the universe, only to be sure that it keeps its consistency under all circumstances. Let us use the paradox of Schrodinger's cat as an example, which I consider a very strong test. Schrodinger, a famous physicist of the last century, had imagined a circumstance when his cat is both dead and alive, simultaneously. To be and not to be alive, both simultaneously. This is impossible by all natural laws of this world, since no system can exist in two distinct states simultaneously. You either have existence or nonexistence, but not both simultaneously, since this is how existence works, it is boolean. We already know the answer: to be or not to be, but not to be and not to be. However, in the current science, this case is still a paradox, to be and not to be, and only quantum physicists claim to be able to solve it. Only quantum mechanics accepts existence and nonexistence simultaneously, probabilistically, which is not actually accurate, but only consensual. To be and not to be, consensually.

Schrodinger has imagined a black box, the perfect confinement, where he places his cat along with a bottle containing poison, and a hammer ready to smash it. There is also a device in the box designed to activate the hammer at a random moment in time, based on radioactive decay. The device activates the hammer randomly, the hammer smashes the bottle, this releases the poison, and the poison kills the cat. Poor cat. While quantum mechanics claims that the cat is both dead and alive simultaneously while in the box, probabilistically.

Yet we have already found that every event, circumstance, object, and subject is unique, since accuracy itself is unique. Could it be different? Could accuracy not be unique? Could the cat be both dead and alive, simultaneously? If this is the case, then our model of the universe malfunctions.

Yet we have always found dualities in the current consensual human society, not through actual dual accuracy, since accuracy is always unique, but through consensus itself, since people can agree on everything inaccurate to be true. While all jurisdictions and ideologies have their own statutes and sets of beliefs where they define as valid anything they please, while these are not accurate, but only valid, since these people validate them themselves, consensually.

Right now in our mental model of the human reality, we have arrived at the point where we could be able to define accuracy itself in any manner we want, not only through abstract consensus, but through the natural laws of this world, which is extraordinary. To be and not to be, defined explicitly and objectively directly by us, directly through the laws of physics itself, through the laws of quantum mechanics. Yet is it actually the case? Let us see.

Let us run our model of reality directly on Schrodinger and his cat. You can do the experiment yourself through your inner self within your inner replica of the world, inner replica of the world freshly updated with this entire model of the human reality, and it is called mental modelling. Ready?

We place the cat and the poison in the black box, along with the device meant to smash the bottle at any random time, we close the box, we start the timer, and we wait. Yet is the cat dead or alive inside the box? We do not know, since the box is a black box, perfectly sealed, as the walls do not allow us to hear the hammer smashing the bottle, nor to see the cat dead or alive. We cannot know of what happens inside the box, if the hammer had already smashed the bottle and poisoned the cat or not, while ignorance hurts, so we just wait. There is no way to tell if the device had been activated, since it is programmed to trigger the hammer at a random moment in time.

Here we are, looking at the perfectly closed black box on the table in front of us, not knowing if the cat is dead or alive inside. Is the cat dead or alive in there? We do not know, with quantum mechanics stating clearly that the cat is both dead and alive simultaneously as we wait. Yet is this actually the case? So we wait some more, with the cat both dead and alive inside the box continuously, according to Schrodinger himself. To be and not to be, indeed, while it was supposed to be the other way, to be or not to be, because accuracy is unique, while the existence defining it is Boolean, having only two states, the existent or the nonexistent. To be or not to be.

Schrodinger's solution to his paradox is that his cat is both dead and alive continuously, which is impossible, since no system can exist in two distinct states simultaneously, because accuracy is unique. Schrodinger had used this imaginary case to prove not only that systems, things, models, and beings can exist in two distinct states simultaneously, by using laws of statistics and probability, but entire elementary particles can exist in two or more different places simultaneously, by having different probabilities of existence associated with each place of existence, or associated with each different state.

Furthermore, Schrodinger stated that, if one single photon of information leaks out of the box to inform us of the state of the cat, then that single photon would have interfered with the experiment, and this could have been sufficient to modify the faith of the cat, by interfering with the random generator device from the box itself, while changing the triggering moment in time. This further implies that once we open the box to check on the cat, we interfere with the experiment, so we never know if the cat was supposed to be dead or alive

The Human Reality

when we checked, when we found it dead or alive in the box, if we had not opened the box at that specific moment in time, but at any time sooner or later. To be and not to be, indeed. What can we do? We mental model some more, in order to find the truth, repeatedly, yet we have an entire model of the human reality at our disposal, so let us use it. Can we use the big bang theory instead?

Let us use our intelligent mental model of a reality, to solve the Schrodinger paradox. We can do so because, as stated in the beginning of this book, we are seeking a model of this world capable to explain everything happening in this world, in society, nature, space, galaxies, edges of the universe, beginning of the universe, future, everywhere, and continuously. While as you notice, the model of our reality has been strong enough so far, helping us model other realities, inner, cognitive, artificial, consensual, social, higher, and outer, along with adjacent concepts, as existence, life, minds, the Deity, connectivity, cognition, our Creator, intelligence, continuum, birth of the universe, classconscious intelligences, meanings of Life, intelligences, and living beings. Therefore, since this black box with the cat inside are part of the universe as imagined by Schrodinger, our model of reality should solve it very easily, even at once, so let us see.

We notice that this perfect box, the black box, does not allow sound or light to escape outside the box, no information whatsoever, which is impossible in real life. Yet we can still consider this case as a theoretical example. Because if there is no communication with the inside of the box, no exchange of mass, information, or radiation, if we cannot access the spacetime continuum inside the box, then the box itself is totally cut out of our universe, which means that it is not part of our universe anymore. This further means that nothing inside the box exists for us, neither objectively nor subjectively. It is worse than Mario, since Mario at least was artificially subjective, and Existence was still able to define it as we played the videogame, through us.

Because now, even Existence itself, which is the existence

of Life, of the wider world, and of our world, cannot define anything within the box, including the state of the cat, because it is a black box. We can certainly assume and imagine anything happening in there, having the same zero probability, since nothing exists inside, as long as the box is closed. Zero probability for the cat to be either dead or alive while in the box. Furthermore, the words 'in there' have no meaning for us anymore, with the box a black box, cut out entirely from our reality. This is the answer for any black box: zero probability of anything within.

Therefore, to answer Schrodinger's question right now, if the cat is dead or alive inside the closed box, we can state clearly that the cat does not exist at all for us, and then everything is possible once we open the box and allow its containment to become part of our universe again. Furthermore, none of the probabilities calculated here in this world and associated with any event taking place inside the black box is accurate, and should never be accepted as accurate, because the content of the black box does not exist at all, neither objectively nor subjectively.

As you study closely Schrodinger and this entire paradox from a wider perspective, you notice an entire quantum physics based just on these types of probabilities assumed for events and subjects that should not apply, because these do not exist at all, in neither assumed initial states. The cat is neither dead nor alive in the black box imagined by Schrodinger, because the cat itself cannot exist for us within a black box, and therefore it cannot be dead or alive simultaneously, since it cannot be anything at all. Once this duality principle fails, the entire quantum mechanics fails. Yet quantum mechanics never feeds the world, since it is always worthless, full of theories. Even the first nuclear bombs were not built by using quantum mechanics.

Quantum mechanics itself is not a natural law of this universe but it is only invented consensually, never applying in the real world. You cannot have anything from this objective world in two or more places and states simultaneously, not even probabilistically. Quantum mechanics should never be part of physics, since physics is the scientific domain studying directly this objective world and all the worlds above correspondent with this world. To be or not to be, and this is the answer.

In the current science, there are millions of scientists spending years of research on this subject alone, time that they could have used doing everything necessary for the world, mostly since they are scientists and the world needs new, accurate, pertinent technology meant to aid with the fulfillment of all human needs. What exactly can society and this world offer to our higher beings and to our higher realities, if our reasoning is so erroneous down here, and if our behavior is so irrelevant and so irresponsible? How exactly do these people serve the Deity through their reasoning and behavior, being so selfish and careless? Hopefully, our world is a joyful collective daydream reality up there, because if it is a collective intelligent reasoning reality, then they are always erroneous. Quantum mechanics.

Can we spend some more time on Schrodinger's cat, to make it better? Because as noticed, you cannot make an actual black box in this world, because you end up taking its contents right out of this world, into nonexistence itself, which is impossible, since you cannot exit this world objectively, because it contains everything that exists objectively, indefinitely. Perfect black boxes are impossible.

What if we do not use a black box, but a normal box instead, a perfectly insulated box more precisely, and still part of this world? Can we still solve the paradox now? Yes, certainly, since the same existence and the same continuum are in the box as they are in the outside world, holding consistency with all the laws of physics and mathematics everywhere. Therefore, just by placing the cat and the poison either inside a box or on the table, on the sofa, or in the garage, it makes no difference, because the laws of physics and mathematics are invariant everywhere. Which means that, this is not a paradox anymore, but normal life, and you can simply apply the common laws of physics and mathematics to solve it.

You cannot apply quantum mechanics anymore, but normal classical physics. Yet since Schrodinger himself was a quantum physicist, he forced everything on his behalf, because at a closer study, we notice how Schrodinger worked hard to apply this paradox to quantum mechanics unnecessarily. First, he used a black box, which is impossible in real world. Then he made the random generator device based on nuclear decay, only for this paradox to become part of modern physics, which includes quantum mechanics. Any random generator can give similar results in this experiment, which means that this is not a modern physics or a quantum mechanics circumstance, but normal, real world circumstance, abiding by the normal laws of classical physics. Which means that this entire paradox invented by Schrodinger is never about the cat being alive or not in the black box, but about this entire paradox being part or not of quantum mechanics, while making quantum mechanics itself not a natural law of this world, but only a consensual theory or scientific belief.

Just by opening the box, don't we interfere with the random generator deciding the faith of the cat itself? No, because as stated, the laws of physics and mathematics are unique, invariant throughout the spacetime continuum, always remaining outside your own lines and lifelines of causality, as long as you do not smash the bottle yourself, or as long as you do not temper directly with the random generator yourself.

To be or not to be, case closed. Let us now test our mental model of reality on the big bang theory itself, one of the most complex models in the world. We have done so repeatedly throughout the book, while now we only enhance our model of the human reality.

We are going to state the major characteristics of the big bang theory, and then we are going to use our model of a reality to test them. Remember, our model of a reality is not finished yet, but it does not have to be finished in order to work. This is not such a tedious test the way it seems, since the big bang cosmological model was poorly defined from the beginning, and it can barely hold consistency on its own, while it is continuously sustained, reinterpreted, and patched up consensually. Furthermore, it is hard to find a real circumstance where the big bang theory had kept its consistency, and this is why it is never used in real life. Imagine using the big bang model on Schrodinger's paradox. You cannot.

Here are the most important characteristics of the big bang cosmological model:

The universe expands.

The big bang took place long ago, when all matter, radiation, and spacetime were condensed in an infinitesimally small point.

The universe is 13.8 billion years old.

These are the three major characteristics of the big bang theory. If our model of the universe proves wrong only one of them, if the big bang model cannot hold consistency by its basic definition and characteristics, then it is clearly erroneous, and it should have no place in science. We have already proven erroneous the redshift, dark energy, dark matter, and inflation, while right now, there is not too much left to help the big bang cosmological theory.

The universe expands. Can this be true? The process of expansion implies an increase in the distance to an object, as a distant galaxy, in rapport with a measuring device or a system of coordinates. The idea of an expanding universe had been forced on us through the speculation that distant galaxies are speeding away from us at relativistic velocities, which was assumed erroneously from the fact that light coming from these galaxies is redshifted. We had already seen how the detected redshift could have other causes, and not necessarily a very high recession velocity.

The simple experiment that we are going to perform is to measure the distance from us to those very distant galaxies and see if it changes in time. If the distance increases, then those galaxies are speeding away from us, and therefore the universe expands. Yet since this happens with all very distant galaxies, we can presume that the universe itself expands at a higher or lower rate. It is obvious that we cannot place a real meter stick from here to a distant galaxy, since it is impossible. This is why we create mental models of the universe, to use them in these impossible circumstances. In our mental model of the big bang, we can even use meter sticks if we want, or not. Let us see.

Those very distant galaxies exist in normal space, we live in normal space, the distant galaxies speed away from us in the same normal space, so it is logic to place our meter stick or system of coordinates in the normal space, in this world. Why do I mention this? Because Einstein himself knew of this simple experiment that we are performing right now, so he found a trick to avoid it. He switched from the normal space to an imaginary or theoretical space, which he called comoving space, while this patched up the big bang once again, in an erroneous manner.

Was it accurate? No. The entire big bang idea had started from the observed redshift of distant galaxies, which implied the fact that distant galaxies move away from us at great speeds in rapport with the real spacetime continuum but not in rapport with any abstract comoving space, which further implied that the universe expands at great speeds. This movement of expansion or recession happens in real space, if the universal expansion is real, and therefore we must measure everything in the normal space, exactly where our real world is. Yet for Einstein to change from normal space to the imaginary comoving space only to stop the big bang model from falling apart, it questions the movement of recession itself, along with the explanation of the redshift. Furthermore, if there is no movement in the real space anymore, no recession, this compromises the supposition of having an expanding universe along with the entire big bang. Case closed.

Einstein had created this duality himself, that the universe expands yet it does not expand, both simultaneously. Duality. He created it out in the open, stating it clearly in his general theory of relativity. Because his general theory of relativity did not work, and he had to patch it up in this manner, which had resulted in patching up the entire big bang theory, in a scientific conflict of interests.

Why the patch and why the lie? The truth is that it is hard to find a stable accurate model of the universe besides an infinite, static, eternal universe, while even that collapses in itself, from the gravitation that the entire universe has and exerts upon itself. While our model of the universe does not collapse, since it uses the law of relativity, holding it in place.

This is what the invisible kingdom does not want, a stable, static, infinite in time model of the universe, since this implies creation, and therefore a creator. Which is actually true, since all living beings and intelligences are creators themselves throughout Life, as they have to experience, perceive, learn, think, and reason within their own created inner cognitive worlds and realities. You have one created inner reality too as a conscious intelligence, which is your own conscious inner replica of the world spanning the cortex.

Creators exist, with one or more for each real, natural, living world and reality, and these count in zillions. I do not state here that Creationism is true or false, as it is part of various ideologies. While all ideologies are consensual, only of the first consensual ideological level, standing below Life and the real world altogether, while remaining incompatible with all living human beings at their third intelligent developmental level. The current science is also consensual, matching all ideologies, including religious ideologies, as it is based on its own scientific consensus, which is also dead, consensually dead, remaining incompatible with the living human beings. When something is incompatible with you, your family, society, civilization, world, and reality, it stands in your way, while harming you and everybody else. Just study all the dark ages closely, to see it yourself.

We notice how the current science is not meant to provide all possible scientific knowledge to all human beings, but it has very harmful agendas, and this is how people die by the millions. Because, if it ever uses the real model of the universe, the invisible kingdom loses its scientific ideology, to let religious ideologies or social ideologies take over the Western Civilization, once again. This is how they lose this world to anyone strong enough to implement his own ideologies in the world, as the old aristocracy and royalty used religious ideologies, or as the dictators of the East use social and political ideologies.

The more Einstein developed his expanding model of the universe, even while he worked on his general theory of relativity, the more he ran into these problems, since none of his equations worked, because all these are erroneous. This is why they invented inflation, since expansion was not enough, and the universe could not expand, even theoretically. Lies always require more lies, so they had to add more lies, and they charged the world billions of dollars for each lie that they added.

Why didn't scientists react? Because they were all in the consensus. Why didn't the scientists outside the consensus react? They probably did, and they still do, but these scientists are not even considered scientists, and they are generally ignored. Why didn't regular people react? After all, this is the very basic knowledge about the world, which is distorted in every manner by science. There is a stereotype in the world to consider that the more complicated knowledge is, and the more you cannot understand it, then the more accurate it is, and therefore the more you must appreciate it, cherish it, love it, and acclaim it, along with the genius scientist inventing it. This is why people love Einstein, they love big bang, and they love quantum mechanics. Furthermore, people had what it seemed to be an ultimate choice to make. It was either big bang theory, or they chose religion and went back to the dark ages, since the old European aristocracy was keeping them there indefinitely. Therefore, people embraced the big bang theory along with Einstein himself, their new savior.

Yet this is not entirely true, because people were getting out of the dark ages anyway, even before the invisible kingdom had arrived, slowly but steadily, and they were moving towards the way life used to be in the last age before religion, in the Age of Aries, now censored by history and science. Which means that you cannot actually choose between aristocracy, invisible kingdom, and the dictators of the East in order to have a better world, between the big bang and Creationism, or between communism and capitalism, since this is as choosing between Hitler and Mussolini to make the world a better place. While if you want to make the world a better place, you have to choose between development and underdevelopment, or between the real and the consensual.

Let us continue the testing of the expanding universe. We know that space itself expands in a model of an expanding universe, not only matter, which implies that our meter stick expands alongside the object we measure, showing us no increase in size. It is as measuring the enlargement of an inflating balloon with a ruler that is drawn directly on the balloon, expanding now alongside the surface of the balloon. Does anything expand according with the ruler? No, not at all.

Take a balloon, draw two stars on it, then draw a ruler near them capable to measure the distance between the stars, then inflate the balloon, and then watch the distance between stars getting bigger. Yet is it actually getting bigger according to the ruler that you drew there? No, since the ruler shows the same distance. According to the ruler, there is never an increase in the size of the balloon, and therefore there is no expansion. Which means that you cannot observe redshift on distant galaxies within an expanding universe, while the redshift the you currently observe has other causes, determined by the law of relativity itself.

Furthermore, you have no outside continuum to measure the rate of your expansion, trivially, since there is no outside for any universe or reality. Therefore, universes cannot behave in any manner wholly, they cannot expand, contract, inflate, or deflate, since they have to do so from an outside physical perspective, while there is no outside for any world, universe, or reality. There is no outside for any universe, because universes include within themselves everything defined objectively real, and therefore universes, worlds, and reality cannot expand or contract, since there is no point of reference to define the change. The big bang theory is erroneous.

Einstein knew it, since this idea had interfered with his theory of relativity, so he invented his new trick to repair everything with a change of the coordinate system, the way he had done years before when he invented a simple constant to repair the collapsing of his spatially finite model of a stationary universe. What was his salary exactly?

It seems that Einstein had one collapsing model of the universe after another, and he always repaired them with lies. He invented the new coordinate system to work in an entirely new artificial space, and called it comoving space, to make a little patch in order to avoid and prove wrong exactly the experiment that we made, stating that if space expands along with the universe, then there is no expansion at all, since all measuring devices expand alongside, showing a zero rate of expansion.

There is expansion, but there is no expansion, or this is what Einstein said. Yet with his invented comoving space and comoving system of coordinates, Einstein changed the expansion of the universe to happen not in the normal space, but in his invented comoving space, using his invented comoving system of coordinates. This is a different world altogether, a different, imagined universe. He changed from real space to commoving space, but then he changed from expansion to inflation, furthermore to change from normal energy to dark energy to balance all equations. While it is easy to change from real worlds to imaginary worlds, because mathematics itself is invariant, balancing all your equations anyway. While with an entire scientific consensus at your side, you are actually omnipotent in the world, consensually.

Here is Einstein's little trick from his general theory of relativity:

'General relativity describes spacetime by a metric, which determines the distances that separate nearby points. The points, which can be galaxies, stars, or other objects, are specified using a coordinate chart or

"grid" that is laid down over all spacetime. The cosmological principle implies that the metric should be homogeneous and isotropic on large scales, which uniquely singles out the Friedmann–Lemaître–Robertson–Walker metric. This metric contains a scale factor, which describes how the size of the universe changes with time. This enables a coordinate system to be made, called comoving coordinates.

In this coordinate system, the grid expands along with the universe, and objects that are moving only due to the expansion of the universe remain at fixed points on the grid. While their comoving distance remains constant, the physical distance between two comoving points expands proportionally with the scale factor of the universe. This means that the big bang is not an explosion of matter moving outward to fill an empty universe, but space itself expands with time everywhere and increases the physical distance between two comoving points.'

Therefore, objects in an expanding space cannot be detected as moving away from each other, since the detecting devices expand alongside space. Therefore, the redshift is not caused by recession, therefore the universe does not expand, therefore the universe is static, and therefore there was no big bang. No big bang. Case finally closed.

Could the universe still expand in some imaginary space, as the comoving space? Yes, certainly, since everything can happen in all imaginary spaces, similar to Schrodinger's cat, all having the same zero probability to be real, since imaginary spaces are not part of the real world, and they do not exist objectively for this universe. Furthermore, study this case closely, to see that the universe is static even in rapport with Einstein's comoving grid, and we are wasting our time here, since there is never a big bang. Because instead of truth, we keep finding people's greed to take over the world, since these are the undeveloped people, and this is the world.

Why exactly did Einstein invent this comoving space? Others invented it, since Einstein is not too capable, but he stole everything from others. There is no fixed point in the universe, this was the major discovery of Galileo Galilei, and Einstein praised him for it. What not too many people know about Einstein is that he still believed in the ether. The ether is thought to be a fluid substance found everywhere in space, with no substance, no color, no mass, nothing, having only its presence, as a grid, and therefore being only imaginary in nature, and not part of this world. When asked what is space made of, Einstein answered that it is made of ether. Who needs ether? It is easier to work with ether in theoretical physics, since it gives you a solid point of reference everywhere in space. It is as having a fixed point anywhere in the universe to help you orient yourself, while there are no fixed points in the universe, everything being relative, by the law of relativity. There is no ether.

Yet all physicists believed in ether at the time of Einstein, the entire old scientific consensus. Poincare called it aether. The law of relativity is based on the fact that everything is relative everywhere, there is no fixed point in the universe, there is no fixed system of coordinates, everything is relative to your our point of view, just the way Galileo stated: 'give me a fixed point in the universe, and I can move the Earth.'

Yet all these happened before the World Wars, during the old world order, during the old European aristocracy. The new world order settled in place at the end of the Second World War with the invisible kingdom on top of the world, since it was the invisible kingdom starting, ending, and winning the wars in order to take over the world, while changing the entire paradigm, the entire mode of society, along with the entire scientific consensus. All physicists dropped their concept of ether simultaneously, along with many ideas in science that you will never hear about, accepting new ideas, as the big bang, particle physics, dark matter, neutrinos, and quantum mechanics. While now, with this imaginary comoving space defining the real world, Einstein had successfully reinstated the old concept of ether, right in the new world order, but based only on different premises. It was a precise grid, generated by the precise position of all space objects. It was a universal system of coordinates, only having a different name, not ether, but comoving space, yet being the same thing, the old ether.

This is how Einstein managed to ignore and bypass the law

of relativity altogether, while derailing cosmology and therefore physics this entire time, since this was the agenda. No more law of relativity, but only Einstein's special and general theories of relativity. How exactly does this feed the world? It does not, but it only indoctrinates the world, while it steals from the world, impoverishing the world, and derailing the world, since this is what the invisible kingdom does in the world. While the old aristocracy and the dictators of the East do the same.

Note that we are considering mostly scientific facts from before World War II, from the old world order, because there is nothing worth considering afterwards, during the current world order, everything being irrelevant in understanding the world, everything being made specifically to distract us from the truth. This is the case not only in science, physics, and cosmology, but in all social domains, at all levels, affecting not only the Masses but the entire world, and implicitly the higher worlds above.

Why hiding the ether in the new world order? Because as I see throughout my models involving the field and all intelligences living directly in the field, along with those that only have their roots in the field, which is also your case as a conscious intelligence, all these descend to the spacetime continuum directly. While if you can only find and see the direct relationship between the field and the spacetime continuum, you can certainly find your way to the matrix holding our world, while through it, you find yourself the direct door or window to the higher worlds. Once you find the higher door, you are one step away from regaining your direct link to your own spirituality and therefore to your highconscious, your direct pathway to your own higher abilities, and your way to learn everything in the world. There are no more lies, you assure harmony, meaning, and fulfillment in the world, and the invisible kingdom along with all dictators from the East lose everything, the entire world. Yet this is not exactly the ether of the old centuries, but it is part of the natural law of relativity, as it involves the spacetime continuum of this world. While with the invisible kingdom censoring both the ether and the natural law of relativity, what exactly are they hiding?

We had performed the test for the expansion of the universe only as seen from inside the universe, and we found it to be static, with no expansion. Yet this is the case with all realities, since from an inside comprehensive perspective, all realities remain invariant objectively, for lack of comparison with anything else objective from the outside, since everything objective is already contained within the reality.

What if the universe really expands, and this expansion can be detected only from the exterior, from outside the universe? As stated, nothing exists outside the universe, so you cannot place your meter stick outside the universe, measure it entirely, see how large it becomes and how fast, then calculate the rate of expansion. There is no spacetime continuum outside to hold your meter stick. Your meter stick exists only objectively and only in this universe, and now wherever you take it, it is still part of this universe. Therefore, you can never take your meter stick outside the universe in order to measure its expansion or inflation. Furthermore, you cannot simply change the word from 'expansion' to 'inflation' in order to make it possible, because you cannot change concepts, since they are unique. Truth is unique. There are other realities within the wider world, yet they are not 'outside' our world, but they are held by other higher realities or by our higher reality as it holds this world. It is the same with videogame realities and mind realities, the spacetime continuum of our universe has nothing in common with them, since our continuum is only in our world, not going into mind realities and videogame realities.

Take a ruler to measure Mario on the hard drive of your laptop in his subjective videogame world and you cannot, because the continuum of this world is different from the continuum of the digital subjective world of Mario. Yet what if Mario expands, since it does so when it eats the mushroom, can you see it expanding on your hard drive too? No. Can you see this world expanding in our upper reality? No, because even from the perspective of upper realities, you cannot define

The Human Reality

the actual size, shape, and comprehensive objective behavior of our universe, since continuums are never linked objectively from one reality to another.

Nothing exists outside any universe from any perspective, because universes have no outside. However, only you can exist 'outside' universes, only through your multidimensional characteristic, and only if these are your own other realities. Right now, your physical body exists outside your mind realities, yet notice how you are not outside realities in this manner, but you are in your other realities directly, through you. Because those are your other realities, as your inner replica of the world where you are as your inner self, the outside world where you are the physical body, and our higher reality where you are your soul. You are always in these other realities, because you already have a self of yours there, being part of your own lifeline of existence. While universes have no outside, and you cannot even go from one universe to another objectively, but you switch from one universe to another or from one reality to another through you, from one self to another, since you cannot travel outside realities objectively with the same self or body. The most common way to do so is to close your eyes, and you find yourself in your inner mind world, in your inner replica of the world, within anything that you can imagine there.

I use the terms 'universe,' 'world,' and 'reality' to define the same concept. The difference between universe and reality is that, while the term universe has been introduced empirically to define everything that exists objectively, the term reality is introduced empirically to define everything that is real objectively. It is the same, since the term 'reality' is synonymous with 'existence' in many instances. Our world can exist in our higher reality, but only subjectively.

How can you go to the higher world in person? You have to go as your soul but not in person, while you have one self in each one of your other realities, and these are highly treasured. While an even more treasured value is you, whenever anyone manages to exploit you in any manner, even without your knowledge.

This world is not up there objectively, to see it inflating, deflating, or doing anything else, because our world is held in an abstract matrix up there, in a larger mind, in a larger computer, or in a larger common mind assisted technologically in every manner, while the space time continuum of the higher world does not apply directly to this world, to allow you to measure this world as a whole.

There is no space outside this world to place your meter stick, not even the space of the higher world. All space is inside the universe, since the universe exists as far as the spacetime continuum lasts, and as long as time lasts. If you still manage to find some space outside the universe to place your measuring stick, then you will notice that nothing happens, since your measuring device is still within the universe, because the space that you find is still in our world. The space that you thought that was outside the universe is still within the universe, and your meter stick still expands alongside the universe if the universe happens to expand, giving you a zero rate of expansion, again. Similarly, you cannot use a meter stick to measure the cars and houses of your dreams along with the entire dream world that you imagine, unless you imagine the meter stick within. It is not the same, since you can imagine your meter stick at any size.

This is the Natural Law of Invariance, we have used it inside and outside the universe, and it states that the universe remains stationary. As stated, our universe cannot even have a shape. Shape itself as seen from the outside has no meaning under these circumstances, since there are no outside space and time allowing its perception. Similarly, the universe has no boundaries in space and time. Boundaries are not defined, since you cannot define them, lacking the meanings to detect them, both from the inside or from the outside.

However, you can define easily everything in this world, including its shape, size, and all its comprehensive characteristics, through higher laws, if you happen to be a soul, yet you must do so from up there, from the higher world. In this higher manner, you make all characteristics of this universe accurate here, in this world, and therefore fixed. You can define everything from here, from this world, through the consensual laws of any jurisdiction or ideology, to make them consensually true in this manner, but only in the specific ideology or jurisdiction that you do so. It is possible not only because everybody does so and it is now part of the consensual human reality, but because jurisdictions and ideologies are distinct, abstract, consensual inner realities. They can be defined by law in this manner, since these laws are higher laws for all inner realities of this world, since jurisdictions and ideologies are abstract inner realities of this world.

Now you can distinguish the meaning or concept of the term 'fixed,' used in religion and spirituality, as in the Earth being 'fixed' on the firmament, meaning that the Earth is set in this manner by higher laws and it cannot be changed from here from this world, since it is now part of the higher laws defining this world. All accurate truth is directly related to higher laws, since they are defined in this manner by higher laws throughout the formation and maintenance of this world in our higher reality.

Does this mean that we will never know whether the universe expands, contracts, or it is stationary? It is actually funny to ask so, because by the natural law of Invariance, considered as a whole, the universe is invariant in every manner, including its shape, structure, and behavior as seen from within. We have seen that the universe is stationary even when testing the supposedly expanding big bang, as it was measured from 'inside' the universe. From the 'outside', our universe can change its shape in any way, it can expand, contract, or remain stationary. However its shape modifies, space modifies accordingly inside the universe, alongside any system of coordinates placed anywhere within the universe. Which makes any outside modification of the universe as a whole undetectable from the interior, since all our measuring devices change shape and size along with space and along with the entire universe. Therefore, nothing is noticed, nothing is detected, nothing is measured, no distortion is detected, there is no change in any dimension on any axis, including time. This is the natural law of Invariance. Not only that our system of coordinates and measuring devices change according to every external modification of the universe, but our own biological receptors and senses of perception are distorted accordingly, along with the shapes of our bodies, so we never notice any change under any circumstance.

Besides, universes are not held wholly by their matrices as these are held by upper realities, but universes are held by characteristics, behavior, and development, in the manner that they are formed, held, and maintained by their matrices. It is the same with all computer worlds, it is the same with all mind worlds, and it is the same with our world as it is held by our higher reality. Therefore, it is impossible to state that universes inflate or deflate as a whole, because universes cannot be defined accurately from the inside or 'outside,' but only directly from their higher realities.

This is how the shape of Earth itself remains irrelevant if it happens that this world is as large as Earth itself, which is mostly the case if this world is formed or created. Therefore, Earth cannot be flat or spherical, because it has no shape as it is held directly by our higher reality. Therefore, nothing can exist beyond the lower orbit of Earth, if this world is as large as Earth itself. Because this world can be the size of Earth, Solar System, entire galaxy, or entire universe that we see from here from Earth. It can come in any size, while it can be defined through higher laws enforcing that you should never be able to identify the actual size, or that the actual size changes continuously with humanity's need for space, it can be anything. This world can be even smaller than Earth, as large as a nation, or a city, wherever you live, and it just happens that you never leave the city, for various reasons, while always assuming that your world is as large as the entire universe, while there is nothing beyond your city. The Earth can also change shape as you move or travel from one city to another, since many dream worlds are in this manner. While if you cannot travel to specific countries or continents to visit or live there, then this might be the case because these places do not even exist, but they are only illusory, used only for decoration, or only to make the world seem larger, to seem complete.

Note that we already knew that nothing exists outside the universe, from the Natural Law of Existence. I only made the model to accept an 'outside' presumption to see if it holds. Since even when I make the model accept exterior modifications in the shape of the universe, as expansion or contraction, the model gives us the results that we perceive daily, all the modifications of the universe as a whole, being anything from a little tilt, bump, shake or scratch, to a massive expansion, contraction, or even a drastic explosion as the big bang. These are undetectable inside the universe, since the entire universe modifies consequently along with all its components, and therefore all detecting devices display no change.

The second major characteristic of the big bang theory was stated by Lemaître years before Hubble, that the big bang was an infinitesimally small point of infinite mass, with the entire universe condensed in it, including all mass, space, and time. This statement is proved wrong by the law of relativity, since you cannot have an infinite mass or infinite density in the universe, since the speed of light is constant, which makes the limit of mass density within elementary particles constant. There is a limit of mass density, equal to the mass density of protons, neutron stars, and pulsars. This is why you have about the same mass density for protons and for neutron stars and pulsars, which is the highest mass density that you can ever reach in the universe, because the speed of light is constant. There are no black holes in the universe either, regardless of what the scientific consensus found to be black holes, because there is an upper limit of the value of mass density and energy density in the universe. However, the big bang is a relativistic circumstance, since that infinitesimally small point which was the universe at the time of the big bang contained all mass, space, and time within it, so its mass and size measured from the outside and from the inside made no sense and they could have held any value, finite, or infinite.

If nothing exists 'outside' the universe, if even when considering its outer shape it makes no sense, then does our universe exist at all? The answer is yes and no. The universe exists for us, objectively, as observed from the inside of the universe, and the universe does not exist objectively as seen from the outside or from any other reality. Yet it can exist subjectively and highjectively, but this has to be the case only through you. Right now, you hold a zillion inner realities within your mind, while you are part of a zillion higher realities above you. You are these through you and through your own lifeline of existence, and this is how you interact through all these realities, through you and through your other selves and avatars, higher and lower.

As a remark, you cannot have a whole universe condensed in a point, since you do not have the means to define that point outside that point. You need an entire upper reality to be able to define that point. Even then, you define that point from within the higher reality, but never from an 'outside.' Therefore, you cannot have a coordinate system to observe the big bang explosion and its expansion or inflation. You have to observe it from the inside, since this is the only objective coordinate system that you have. While everything is invariant inside when considered comprehensively for the entire spacetime continuum, since you have no point left to place your coordinate system to observe the big bang, and therefore again, there was never a big bang, and will never be one, at least not naturally, but only consensually.

You already notice that there are more important matters to consider while studying the human reality, as the correspondence with all higher and lower realities involved, and how these affect everything here in this world, influencing the human reality.

How can you tell exactly which knowledge is accurate? Again, through higher laws and natural laws you can always be certain, since you define it that way. More precisely, if you have knowledge of the specific higher laws used in forming and maintaining this world, you can be sure that they correspond to accurate knowledge here on Earth, by default. If you do, you are capable to know everything on Earth, past and present. Yet you might have to be a higher being to do so.

Until then, you have to use only the natural laws of the universe along with your own reasoning at your best, in order to be able to judge on your own what is accurate and what is not, and this makes the difference between a burden and a successful life. While as we always notice, inaccurate knowledge is of the first level, and it can come in form of corrupted, erroneous, inaccurate knowledge, or in form of any consensual knowledge, as beliefs, juridical laws, or entire ideologies, stereotypes, and jurisdictions.

Why do many religions and schools of thought claim to have higher knowledge, while calling themselves ideologies? It is the same with the current consensual science claiming to be the scientific domain of the human society while still claiming to use a scientific consensus and not accurate knowledge about this world, while calling all its scientific knowledge theories. Because you cannot lie to the living human beings, or at least not at the significant level affecting the human knowledge and the human development on Earth, or you face higher consequences. Therefore, you have to tell them upfront that you are lying to them, that it is only a game, a play, and act, through every legal scheme possible, and this is what the invisible kingdom does. While the dictators of the East can do everything that they want to you, since once you vote for them in unanimity, as they are the only ones on the list, you give them your consent to do everything to you, and so you die. Which is the case with all living beings of Life, as already seen, all dying even in mass whenever they fail to reach accuracy, meaning, and therefore fulfillment.

Which exactly are the natural laws of the universe? We already know some, while various ideologies mention others, which can be accurate or not. Therefore, we must always be very careful with the knowledge already available about the higher laws and the natural laws of the universe fabricated on Earth. There are many ideologies stating the natural laws of the universe, and therefore the names that I use can be consistent or not with what you already know. Furthermore, I include natural laws of the universe generated by all our higher realities, and not only by our higher reality holding our world.

Note that the higher laws are different than how we end up interpreting them here in this world. We cannot have them accurately here but we are forced to interpret them, because you cannot transfer objects and subjects from one reality to another directly and accurately, but only their interpreted and even speculated copies. Therefore, our natural laws of the universe can be interconnected or not in the manner described here as they do in the higher worlds. This is the natural case, because as always seen, there are human beings and souls capable to define on their own everything happening in this world by higher laws, and in this manner, these higher beings are capable to bring higher knowledge on Earth, just the way Prometheus had brought down the knowledge of fire from above.

How big is our universe? The universe is as large as we can see or detect, which is as far as there is still a time rate. Once light stops coming, it means that we are too far for them to reach us, and we are beyond their event horizon. The scientific consensus already states that there are galaxies as far as a specific limit, with nothing beyond. Even though we still have the means to detect farther, there is nothing beyond, just 'empty space,' the way they had stated. That is the furthest limit of our universe, the surface of the largest sphere there can be, which is our event horizon.

Can you travel too far and too fast, to go out of the universe? No. The reason why our universe is enclosed by our event horizon is because nothing can reach us from beyond, not even light. Simultaneously, light emitted by our star and galaxy cannot escape our universe, so it is always trapped here. Since according to relativity, light can travel an infinite space, regardless of how dim it becomes, while this is the case because relativity does not agree with quantum mechanics. Einstein did not agree with Bohr personally, along with his entire quantum mechanics, while Einstein was deciding the faith of physics. According to quantum mechanics, light is quantized, it should have a limit in its amplitude, and therefore it should not travel infinitely, but only until it reaches a certain quanta. This was the explanation coming from quantum mechanics. In our model of the universe, time is zero outside our universe, and therefore light reaches a zero frequency there, and does not exist anymore, since light goes as far as the event horizon, which is the definition of the event horizon. You cannot go out of the universe, as seen from home.

Why is the universe stable? Why does it not collapse in the center of the universe, exactly where we are, the way it had happened with Einstein's first model of the universe? Because each star, each galaxy is a center of the universe, and therefore it is in equilibrium, being attracted with the same force by all galaxies around it. Everything is in equilibrium, and therefore the universe is static. I had to give this explanation in order to be easier understood, yet the answer is the other way around, everything is in equilibrium since all centers of mass are stable, set in place by the entire universe. This is how everything is in equilibrium, and this is how the universe is static and undisturbed, endlessly.

Could we detect or derive now the true distortion created by the mass of the entire universe? Because this is very important in defining the stage of our civilization. Is it a distortion in space created by the mass of all galaxies, by the mass of the universe, or is it a distortion in time? Because, if it is a distortion in space, then the universe seems to be expanding, even though it is a simple curvature of space, the universe being actually static. If it is a distortion in time, only time slows down significantly, as it is distorted by the mass of the entire universe, and the universe is static. Which is which? The answer is that it is a distortion in time, since galaxies are not moving anywhere, since they cannot, because you cannot displace centers of mass, and therefore you might not be able to displace the field at the very large scale of entire galaxies. Furthermore, distortions in space expand space at the time of these distortions, and then they only keep space distorted. Distortions in space or curvatures in space only increase distances along the axis of the distortion, creating the famous spaghetti phenomenon from black holes: everything falling in a black hole increases its length towards the center of the black hole.

Note that this distortion is observed from the outside of the distortion, and not by you, the one falling in the black hole, since all your means of observation distort accordingly, by the Natural Law of Invariance. Another interpretation of this temporal distortion is that we place the origin of our system of coordinates in space, normally, as we do in all experiments, since we always live in our spacetime continuum, as we only experience indirectly time passing, using a stopwatch.

We live in our spacetime continuum, in a physical world. Assuming that the redshift of distant galaxies is caused by a curvature in space is as assuming that the Sun revolves around Earth once a day. This is true most of the time, since everything that we do in life we do in rapport with the surface of Earth. Yet technically, the Earth rotates around its own axis, giving us the illusion that the Sun revolves around Earth, or this is the case if the Sun is not an illusion itself. All curvature, all distortion created by the mass of the universe is done in time, not in space, and it can be detected not only from our normal space, but from the comoving space of Einstein, which is also stationary in our model of the universe. Time is distorted, causing the redshift, while space is always in a normal state.

How does mass distort time? Mass slows down time completely. Pure mass exists mostly in form of protons and sometimes in form of neutron stars and probably neutron galaxies. Everything has the same mass density and it is in eternal equilibrium since their time is zero. All objects around are formed by atoms, with all protons condensed in the nucleus. As a whole, there is significantly more empty space than nuclear space, and this gives to normal objects the usual low density that you know. However, all protons, neutrons, and nuclei stop time completely for the little space that they occupy and in its immediate vicinity. You will never notice the slowing down of time caused by an apple, yet all mass in the universe manages to slow down your time all the way to zero. From your perspective, your clock always ticks normally, yet from the perspective of a distant galaxy, Milky Way shows a redshift, since time is at a lower rate here, according to them. Because all mass of the universe slows down time here, even all the way to zero, depending on how far they are from us, since for more distant galaxies, redshift is more substantial. For the most distant galaxy that we can detect, our redshift will be so pronounced, our light will be so dim, and our rate of time will be so low, that we barely exist for them. For them, we mark the limit of their universe, while for us, they mark the limit of our universe, since the exact redshift that they observe on us, we observe on them, as the universe is homogenous.

Now we know the size of our own universe, since it spans as far as the time rate is above zero, which is about as far as the current astronomers are capable to detect galaxies. This relativistic phenomenon taking place at very large distances is capable to maintain the universe static, without collapsing in itself.

Now we understand how our universe is limited in space, not by the end of space or the end of the electromagnetic field, but by the distortion of time. Note that, if I would have considered the relativistic curvature to be not in time but in space, then our universe was considered infinite at our event horizon, which would have made everything more complicated, since it would have been harder to understand the universe.

It is always the same natural law of relativity at play, since you can always state that your car moves forward, or that the entire world moves backward, it is your choice. Yet you have to remain consistent with what the entire world perceives, that you move forward and not that they move backward, because they can assure you that they do not go anywhere.

How does mass stop time? If the universe is infinite, this means that our time should be distorted infinitely by the mass of the entire universe, so our clocks should not move at all as seen from the very distant parts of the universe. This can be true, since there is a limit in mass causing total distortion.

This is the case with each proton. Their rate of time is zero, this is why they are infinitely stable particles, and they do not have to decay to a black hole in order to reach this state of equilibrium. We can state that mass at nuclear level exists in a zero rate of time. It cannot increase its energy by any means, and when it does, the extra energy is immediately converted into radiation. Fire is a common example, or stars, or particles under relativistic Cherenkov circumstances.

Our redshift signals a breaking down of the Natural Law of Conservation. Where is the energy lost by distant galaxies signaled by the redshift? We have seen how the mass of the universe slows down time. A slower rate of time means a lower action, while we know that action is conserved. Where is the difference? The answer is always the same: the field is converted from one state to another. In this case, the field is converted from electromagnetic radiation to matter, or from photons to protons. Because not exactly energy and momentum are conserved, but energy density and momentum density in the spacetime continuum are conserved, causing all redshift and plasma in the world.

How is light converted into protons? It is the simple Cherenkov radiation on a larger scale, because at very large scales, you are always in relativistic circumstances, and you must use the law or relativity. We know that the speed of light in vacuum is invariant in any system of reference. However, that is true only when time is unchanged. When time is slowed down by large energy density as by large mass density or by large speed, it seeks to make light change its speed to a higher value relatively but it cannot, since its speed remains constant in all systems of reference. Light does not only bend or chance direction as it does during normal normal optical refraction, but it shifts the field involved to another state, from electromagnetic radiation to matter. When this phenomenon is local, as it happens when gamma radiation reaches Earth's atmosphere, you obtain pairs of particles-antiparticles, which is basic Cherenkov radiation. Yet when this phenomenon is universal, when Cherenkov radiation comes from the entire universe, you obtain one proton. While currently, around the Solar System, there are about one hydrogen atom for each cubic inch.

Life and Intelligence always decide how matter is transformed to electromagnetic radiation, and how electromagnetic radiation is transformed to matter, to protons.

Note that this proton is not only created, but it has to be maintained there by the entire universe in order to exist, along with its own center of mass. The proton does not have to remain there since it can move around locally, while obeying the conservation of momentum and the conservation of energy. However, its center of gravity will always be where it had been created, and this is how it influences the universe in a similar manner from wherever it goes, endlessly. The universe remains stable in this manner, since electromagnetic radiation creates matter, and matter creates electromagnetic radiation, in a continuous shift of the field from one state to another, in perfect agreement with all natural laws of the universe.

This is the human reality, and we have just seen it presented in all domains and through all perspectives. You do not live your life only within this world, but you go to other worlds and realities every time you reason, dream, daydream, worship, and project. Yet you do not even live in the real world if you are not careful, because you end up consensual, throughout the multitude of jurisdictions, hierarchies, and ideologies of the current consensual society, as these are consensual realities altogether, part of the consensual human reality. You can end up living only through your strong convictions in your dream world, while that is not part of the human reality either, but part of the human inner mind reality, and it is not the same. These worlds are distinct universes, natural or consensual, they are simple consensual jurisdictions, or they can be entire astral planes of existence. We do not go there as human beings, but as our higher and inner selves, depending if our other realities are higher or lower in nature, while it is always important to identify and study them in all details, for a more meaningful and a more fulfilling existence.

Because you are more than a human being while you are more than what others decide that you are and should be as a human being, in this world and in all your other realities. You are an intelligent living human being, and therefore you are more meaningful and more capable than entire worlds and realities, because you are entire worlds and realities yourself, through yourself and through everybody else that you interconnect with throughout the human reality, at all your levels of existence. Just make it a good one.

The End

This book series continues with the next book, "Astral Planes and Your Other Realities." Here is a short synopsis:
Can you visit other worlds and realities? Can you go to live your life there, in your other worlds and realities? People can always tell beautiful paranormal stories, or who knows, you might have already been there yourself, saw everything, felt everything, and now this is why you search everywhere, in order to learn more about your experience.
Was it real? Are your other worlds real, as real as this world? All realities are real, in a rather trivial manner. Furthermore, all realities are objectively real, but only for as long as you are there, since existence defines realities similarly, determining closely your continuous firsthand experience in each one of them. Furthermore, many realities are part of life, part of this life that we have here in this world. If you have found your way in these other planes of existence in a conscious manner, it might have been a natural process altogether, and not exactly a random experience. It had a

purpose, while you might have missed fulfilling it, and now it is certainly important to know everything that you can ever discover.

Yet if you have never had a paranormal experience yourself, and now you simply wonder what is going on, there are very powerful drugs that you take regularly with your food, drinks, drugs, and medicine, meant to hold you forcefully in this world, and this is what you do, you remain here in this world, continuously. This happens with everybody, or almost, depending on where you live, or depending on your development and genetic background.

There is a difference between astral planes, the natural human environment, and your other realities, because existence is relative, coming in three distinct levels, used to distinguish between your higher and lower realities, while you can also understand all your realities through your mind, reasoning, awareness, and imagination, since even this world makes sense to you only as part of your reasoning, awareness, mind, and imagination, and not exactly directly as anyone might expect.

Because there is no other way to experience anything in life and in the wider world but through your perception, reasoning, and understanding, and through the multitude of your selves, intelligences, and identities present throughout all your worlds and realities. This is why you cannot understand astral planes and your other realities, if you do not understand your cognitive system first, along with your intelligences, selves, memories, meaning, and expectations, since everything is interconnected. This interesting entanglement of meanings, constraints, and expectations causes the ultimate truth of your wider existence to remain hidden beneath strong consensual conditions and within tedious loops of reasoning, remaining inaccessible in this manner to the ignorant, the addict, and the follower of various ideologies, while allowing the truth only to those living life freely, consciously, and in full awareness and understanding of the wider world, through the fulfillment of all natural higher level needs and meanings. Because these are your developmental opportunities that you experience in each one of your worlds and realities, while following the fulfillment of your natural, intelligent human needs for higher experience and higher development.

This book studies you and your life and existence throughout all your realities that you use, encounter, inhabit, create, and co-create throughout your wider existence, helping you understand who you are through all your selves and intelligences, as you live your life throughout all your worlds and realities. If you seek to gain wider understanding of who you truly are, this book is for you.

ABOUT THE AUTHOR

Valentin Leonard Matcas, M.Ed., is a researcher, physicist, mathematician, educator, and an author of nonfiction and fiction books, including the entire "Human" book series. Valentin Leonard Matcas wrote the "Human" book series in the following order: "The Human Needs", "The Human Addictions," "The Hierarchy of Needs," "Stay in Shape, Lead a Healthy Life," "The Human Origins," "The Human Society," "The Human Conspiracy," "The Human Mind," "The Human Reality," "Astral Planes and Your Other Realities," "Life," "The Hierarchy of Intelligences," "The Human Intelligences," "The Human Thoughts," "Mental Models and Successful Ideas," "The Human Attitudes," "The Human Stereotypes," "The Human Ideology," "Modes of Life," "The Human Development," "Patterns of Development," "The Human Lifestyle," "Heal Yourself," "The Human Civilization," "The Human Religion and Spirituality," "The Human Rights," "Higher Laws," "Natural Laws of the Universe," "Existence," "The Human Condition", "Lifelines of Causality," "The Human Behavior," "Flat Earth," "The Human Environment," "The Human Meaning," "The Human Reasoning," "The Human Interconnectivity," "The Consensual Matrix," "The Matrix of Life," and "The Human Knowledge."

Valentin Leonard Matcas writes about terrestrial and alien civilizations, about life in the universe, the way it develops and intertwines across galaxies, about powerful beings as they control and reshape the universe, and about normal living human beings from Earth caught in this beautiful, wider, outstanding interconnectivity. Valentin Leonard Matcas creates a living, warmer universe in his books, teaming with life and vibrancy, on all levels of existence. Valentin Leonard Matcas also wrote "The Storyteller" book series, including "The Storyteller," "Starship Colonial," and "Unlimited," and "The Culling" book series, including "The Culling," "The Dream of the Dead," and "The Last Man on Earth."

When he does not work on his books, Valentin Leonard Matcas enjoys researching, hiking, swimming, kayaking, skiing, snowboarding, biking, reading, listening to music, and playing strategy videogames. You may discover all his books, videos, and articles.

www.ingramcontent.com/pod-product-compliance
Lightning Source LLC
Chambersburg PA
CBHW020627220526
45464CB00001B/51